SpringerBriefs in Molecular Science

Chemistry of Foods

Series editor

Salvatore Parisi, Al-Balqa Applied University, Al-Salt, Jordan

The series Springer Briefs in Molecular Science: Chemistry of Foods presents compact topical volumes in the area of food chemistry. The series has a clear focus on the chemistry and chemical aspects of foods, topics such as the physics or biology of foods are not part of its scope. The Briefs volumes in the series aim at presenting chemical background information or an introduction and clear-cut overview on the chemistry related to specific topics in this area. Typical topics thus include:

- Compound classes in foods—their chemistry and properties with respect to the foods (e.g. sugars, proteins, fats, minerals, …)
- Contaminants and additives in foods—their chemistry and chemical transformations
- Chemical analysis and monitoring of foods
- Chemical transformations in foods, evolution and alterations of chemicals in foods, interactions between food and its packaging materials, chemical aspects of the food production processes
- Chemistry and the food industry—from safety protocols to modern food production

The treated subjects will particularly appeal to professionals and researchers concerned with food chemistry. Many volume topics address professionals and current problems in the food industry, but will also be interesting for readers generally concerned with the chemistry of foods. With the unique format and character of SpringerBriefs (50 to 125 pages), the volumes are compact and easily digestible. Briefs allow authors to present their ideas and readers to absorb them with minimal time investment. Briefs will be published as part of Springer's eBook collection, with millions of users worldwide. In addition, Briefs will be available for individual print and electronic purchase. Briefs are characterized by fast, global electronic dissemination, standard publishing contracts, easy-to-use manuscript preparation and formatting guidelines, and expedited production schedules.

Both solicited and unsolicited manuscripts focusing on food chemistry are considered for publication in this series. Submitted manuscripts will be reviewed and decided by the series editor, Dr. Salvatore Parisi.

To submit a proposal or request further information, please contact Dr. Sofia Costa, Publishing Editor, via sofia.costa@springer.com or Dr. Salvatore Parisi, Book Series Editor, via drparisi@inwind.it or drsalparisi5@gmail.com

More information about this series at http://www.springer.com/series/11853

Suresh D. Sharma · Michele Barone

Dietary Patterns, Food Chemistry and Human Health

 Springer

Suresh D. Sharma
Department of Biochemistry
and Molecular Biology
Pennsylvania State University
University Park
State College, PA, USA

Michele Barone
Associazione "Componiamo il Futuro"
(CO.I.F.)
Palermo, Italy

ISSN 2191-5407   ISSN 2191-5415   (electronic)
SpringerBriefs in Molecular Science
ISSN 2199-689X   ISSN 2199-7209   (electronic)
Chemistry of Foods
ISBN 978-3-030-14653-5   ISBN 978-3-030-14654-2   (eBook)
https://doi.org/10.1007/978-3-030-14654-2

Library of Congress Control Number: 2019933839

This Springer imprint is published by the registered company Springer Nature Switzerland AG
The registered company address is: Gewerbestrasse 11, 6330 Cham, Switzerland

Contents

Chapter 1
Deleterious Consequences of Dietary Advanced Glycation End Products on Human Health Due to Oxidative Stress and Inflammation

Abstract The evolution of modern food productions has progressively modified dietary patterns in the industrialised world with the increase of chronic diseases. The technological progress in food industries might be roughly correlated with the concomitant augment and differentiation of certain diseases. The diversification of the problem has to be expected because foods and beverages imply a multidisciplinary knowledge, including microbiological causes and effects, chemical and physical features, technological factors also linked to the First and the Second Laws of Food Degradation, nutritional and hedonistic behaviours, and public health. The aim of this chapter is to describe one of the most interesting issues correlated to foods normally containing 'Maillard reaction products': the production of advanced glycation products. High levels of these complex compounds are reported in relation with chronic diseases (renal failure, diabetes, oxidative stress, inflammation, etc.). Because of the strict relationship between the observed increase of glycation end products and thermal food processing, the chemical identification and quantification of these compounds in foods are extremely useful. In addition, this chapter described some possible solutions and health guidelines for consumers against deleterious effects represented by advanced glycation products.

Keywords Advanced glycation end product · Aminoguanidine · Chronic disease · Diabetes · Pyridoxamine · Retinopathy · Vitamin C

Abbreviations

AGE	Advanced glycation end product
FAO	Food and Agriculture Organization of the United Nations
GOLD	Glyoxal–lysine dimer
MRP	Maillard reaction product
MOLD	Methylglyoxal–lysine dimer
CML	N-ε-(carboxymethyl)lysine

ROS   Reactive oxygen species
RAGE   Receptor for AGE
WHO   World Health Organization

1.1 Advanced Glycation Products and Health Issues

The evolution of modern food productions has progressively modified dietary patterns in the industrialised world with the demonstrated increase in chronic diseases. In other words, it has been often reported that the technological progress in food industries might be roughly correlated with the concomitant augment and differentiation of certain diseases, on the basis of the last available scientific literature (Devcich et al. 2007; Hong 2016; Lobo et al. 2010; Sanders 1999; Scott 2003; Trowell 1972; Vlassara 2005; WHO/FAO 2003). Interestingly, the correlation between modern food products, on the one side, and the increase in health problems, on the other hand, concern many different aspects. This diversification of the problem has to be expected because foods and beverages imply a multidisciplinary knowledge (Barbera and Gurnari 2017; Brunazzi et al. 2014; Gurnari 2015; Parisi 2002, 2003, 2004; Parisi et al. 2006). The analysis should include at least:

(a) Microbiological causes and effects (because bacteria and other living microorganisms can survive and spread in acceptable conditions, assuming the food or beverage 'support' is good enough.
(b) Chemical and physical features of the product. The packaging material(s) should be considered in this ambit because of its identification with the purchased food product(s).
(c) Technological factors concerning the process. As a simple example, the comminution of certain edible raw materials is potentially able to decrease the 'resistance' of the food products against degradation, with the exception of food products which can be treated, modified, processed, packaged and stored with the main objective of shelf-life extension. On the other hand, observable risks may depend on the First and the Second Laws of Food Degradation (Parisi et al. 2004; Troise et al. 2013).
(d) Other physical reasons
(e) Technological factors related to the processing flow and possible variations, with demonstrable or forecastable effects on the final product(s).
(f) Nutritional behaviour(s) of the 'normal consumer' (Parisi 2012, 2013), and contrasting strategies in terms of public health (Anjali Jain 2004).
(g) Hedonistic concepts related to the advertised food product
(h) Health Sciences, in terms of public safety and correlated food hygiene. The inclusion of nutritional behaviours in this ambit might be sometimes accepted, although normally recommended guidelines do not comprehend the emersion of new safety issues such as microbial spreading by *Escherichia coli*, and other similar situations.

As a consequence, many possible worries could be considered, and the ambit of food safety issues is really broad. The aim of this chapter is to consider one of the most interesting—and emerging—issues correlated to cooked foods and/or all types of foods normally containing 'Maillard reaction products' (MRP), especially if these products represent a well-specified MRP subcategory: advanced glycation end products (AGE). These molecules are extensively studied when speaking of food technology on the one side and of food safety on the other side (Friedman 2003; Fu et al. 1994; Gkogkolou and Böhm 2012; Hayase et al. 2005; Singh et al. 2001; Zhang et al. 2008). In general, it can be affirmed that high levels of these complex compounds have deleterious effects when speaking of chronic diseases and related effects such as renal failure, diabetes, oxidative stress, inflammation, Alzheimer's, and Parkinson's. (Kokkinidou 2013; Ramasamy et al. 2005). Because of the strict relationship between the observed increase of AGE (also in in vivo conditions) and thermal food processing, the chemical identification and quantification of these compounds in foods are extremely useful (Parisi and Luo 2018; Simpson 2012; Singla et al. 2018).

1.2 Classification of AGE by the Chemical Viewpoint

From the chemical point of view, AGEs are strictly linked with the later steps of the Maillard reaction, a complex of reactions commonly observed in processed foods and beverages, but also in in vivo systems. Studied AGE products are also correlated with in vivo studies (Kokkinidou 2013; Parisi and Luo 2018; Simpson 2012); however, it has to be highlighted that thermally processed foods can exhibit notable AGE amounts (ALjahdali and Carbonero 2017; Henle 2008; Lund and Ray 2017; Van Nguyen 2006).

Reducing sugars, on the one side, and 'long-living' molecules such as lipids, proteins, and/or nucleic acids, on the other side, can react via a non-enzymatic process. This complex step, called 'glycation', occurs in the later steps of the Maillard reaction, and exactly after the production of (i) a Schiff's base and (ii) the subsequent Amadori rearrangement with the production of N-glycoside, in this situation it is N-(D-glucose-1-yl)-L-asparagine (Ahmed 2005; Friedman 2003; Parisi and Luo 2018; Paul and Bailey 1996; Singla et al. 2018). After this step, two different AGE products—protein adducts and protein crosslinks—can be obtained irreversibly, while the initial production of Schiff's bases and Amadori products is reversible (Gkogkolou and Böhm 2012).

In detail, the general production of AGE can follow distinct pathways as follows[1] (Gkogkolou and Böhm 2012; Singh et al. 2001; Thorpe and Baynes 2003):

(a) Glucose reacts with an amino compound; the Schiff's base and the subsequent Amadori rearrangement product are obtained. The Schiff's base can be turned into α-oxoaldehydes via rearrangement to fructosamine; oxoaldehydes can

---

[1]Glucose, an example for the reducing sugar, corresponds to the starting point.

subsequently give AGE products. Moreover, the Amadori product can be rearranged again and subsequently give:

(a.1) AGE products via oxidation (e.g. pentosidine)
(a.2) 'Glycoxidation end products' via non-oxidative pathways (e.g. pyrraline)
(a.3) Be turned into α-oxoaldehydes; these molecules can subsequently give AGE products.

(b) Glucose can be also directly degraded to α-oxoaldehydes; these molecules can subsequently give AGE products
(c) Alternatively, glucose can be converted into sorbitol, which subsequently enters another reaction pathway as fructose. This pentose can directly react with lipids via peroxidation, or be converted in other ways to produce α-oxoaldehydes. Anyway, the production of AGE by α-oxoaldehydes is the final step.

Substantially, AGE may be obtained by the reducing sugar reacting with the amino acid in a classical manner (via Schiff's base and/or Amadori rearrangement), or via the sorbitol–fructose pathway. Produced α-oxoaldehydes are normally identified with glyoxal, methylglyoxal, and 3-deoxyglucosone (Edelstein and Brownlee 1992; Parisi and Luo 2018; Singh et al. 2001; Singla et al. 2018). Despite the apparently low number of intermediates able to produce AGE, the number of these final products can be remarkable. Moreover, the heterogeneity of the whole AGE category should be considered (Gkogkolou and Böhm 2012), in spite of their apparent similarity when speaking of visual properties. In fact, AGE—such as many of known MRP—have a distinctive yellow-to-brown colour (John and Lamb 1993; Raj et al. 2000).

The problem of AGE production with the additional production of melanoidins (Parisi and Luo 2018; Simpson 2012; Singla et al. 2018) is strictly correlated with starting 'raw materials', including obviously foods and beverages (Van Nguyen 2006). For this reason, MRP in foods can effectively influence the production of AGE in human beings and animals by (i) increasing the rate of AGE production in consumers and (ii) augmenting the amount of AGE obtained in animal livestock. Also, it has been reported that the production of AGE can be time-dependent and accelerated by transition metals, while the following compounds or extracts may inhibit in vivo AGE production for evident therapeutic or inhibitory purposes (Gkogkolou and Böhm 2012; Rahbar and Figarola 2003):

(a) Reducing agents
(b) 'Nutraceuticals'
(c) Nucleophilic hydrazines
(d) Specific 'AGE breakers' or AGE scavengers
(e) Enzymes against various Maillard reaction intermediates.

On the other side, AGE and melanoidins can be remarkably produced by means of high processing temperatures for food productions (roasting, cooking, sterilisation,...); prolonged storage at high temperatures is also reported to have some influence (Lund and Ray 2017).

Because of the remarkable number of possible AGE, this chapter is dedicated to a selected group of these compounds, with concern to their health importance.

1.3 AGE of Interest by the Health Viewpoint. Possible Solutions and Guidelines for Consumers

The impact of AGE on human health is a debated argument because of the number of positive or negative claimed effects. At present, it has been reported that AGE could positively enhance the antioxidative behaviour of melanoidins, when found in foods (Delgado-Andrade 2014; Lund and Ray 2017). On the other side, AGE production could be correlated with positive effects when speaking of T-cell immunogenicity, pathogenesis of allergenic reactions from foods, and inflammation diseases (Koschinsky et al. 1997; Ilchmann et al. 2010; Toda et al. 2014; Urribarri et al. 2005; Vlassara et al. 2002). In particular, the relationship between AGE formation and inflammation has been described as a repeated cycle explained with the simple model: inflammation → generation of reactive oxygen species (ROS) → enhanced AGE production → enhanced inflammation → enhanced ROS generation, etc. (Traverso et al. 2004). However, these results have been recently questioned because of the lack of information related to their biological role and the reliability of involved studies with concern to immunological procedures and the estimation of AGE in studied foods (Lund and Ray 2017; Tessier et al. 2014). The interest in food allergens and analytical devices able to detect them in foods with reliable results is growing in recent years (Popping et al. 2018); for this reason, AGE's are often researched.

Moreover, AGE compounds have been often considered in relation to ageing. In detail, it has been reported that the accumulation of AGEs in living tissues may favour ageing and modifications of the genetic code, with possible relations between glycation/glycosylation rates related to skin collagen and long-life expectations in mice (Ahmed et al. 1997; Baynes 2002; Kirkland 2002; Odetti et al. 1998; Schleicher et al. 1997).

Some of the most representative AGEs, normally associated with carrier proteins such as albumin in blood, and often found in human skin (collagen above all), are (Chen et al. 2016; Gkogkolou and Böhm 2012; Hellwig et al. 2015; Lund and Ray 2017; Ramasamy et al. 2005; Šebeková et al. 2008; Seiquer et al. 2014; Singh et al. 2001):

- N-ε-(carboxymethyl)lysine (CML), normally considered as a marker of oxidative stress from the metal-catalysed oxidation of polyunsaturated fatty acids and carbohydrates in presence of proteins (Fig. 1.1). It is a non-fluorescent adduct, similarly to pyrraline
- N-ε-(carboxyethyl)lysine
- Pentosidine (Fig. 1.2), a fluorescent glycoxidation product by non-oxidative pathways, it is able to produce protein-protein crosslinks
- Pyrraline
- Methylglyoxal–lysine dimer (MOLD). MOLD and glyoxal–lysine dimer (GOLD) are able to produce non-fluorescent protein-protein crosslinks (Fig. 1.3)
- [5-(5,6-dihydro-4H-pyridin-3-ylidenemethyl)furanyl]-methanol
- Glucosepane
- Fructoselysine.

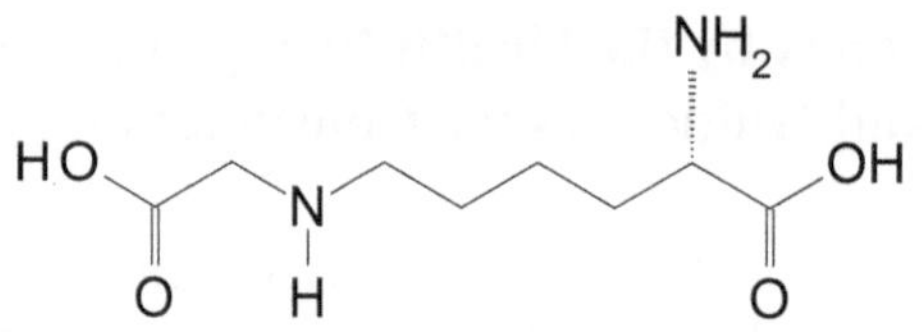

Fig. 1.1 Molecular structure of *N*-ε-(carboxymethyl)lysine (CML), normally considered as a marker of oxidative stress from the metal-catalysed oxidation of polyunsaturated fatty acids (and carbohydrates) in the presence of proteins. This non-fluorescent adduct is one of the most representative AGEs, normally associated with carrier proteins such as albumin in blood, and often found in human skin (collagen above all). BKChem version 0.13.0, 2009 (http://bkchem.zirael.org/index.html) has been used for drawing this structure

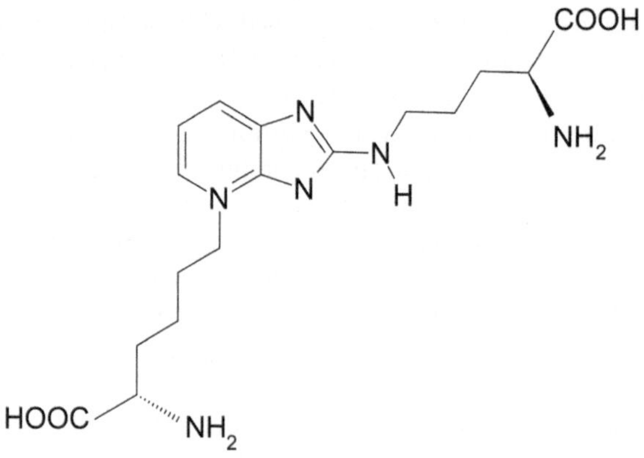

Fig. 1.2 Molecular structure of pentosidine, a fluorescent glycoxidation product. This molecule is able to produce protein-protein crosslinks. BKChem version 0.13.0, 2009 (http://bkchem.zirael.org/index.html) has been used for drawing this structure

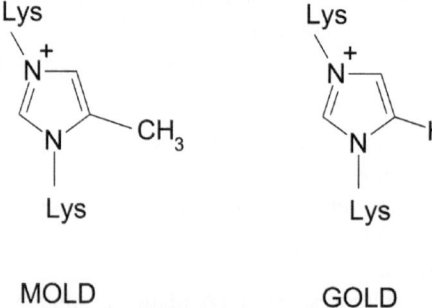

MOLD   GOLD

Fig. 1.3 Molecular structure of methylglyoxal–lysine (MOLD) and glyoxal–lysine dimers (GOLD). These compounds are able to produce non-fluorescent protein-protein crosslinks. The 'Lys' groups are for 'lysine'. BKChem version 0.13.0, 2009 (http://bkchem.zirael.org/index.html) has been used for drawing these structures

These and other AGEs can be of endogenous nature or introduced in the body by food consumption or tobacco use. It should be also considered that their covalent crosslinking nature gives the possibility of long-life structures by association with collagen; on the other hand, short-life proteins are not often interested by crosslink reactions (Gkogkolou and Böhm 2012). Interestingly, available studies have reported that the production of similar crosslinked structures can be correlated with diabetes and aortic dysfunction, especially in aged patients, renal dysfunctions (sclerosis), general capillary thickening, and lipoprotein agglomeration (Dyer et al. 1993; Singh et al. 2001). In addition, AGE can easily interact with specific pattern recognition receptors, named 'receptors for AGE' (RAGE), with other important effects (Gkogkolou and Böhm 2012; Ramasamy et al. 2005). RAGE activation determines the transcription of a remarkable number of proinflammatory genes, with the consequent self-repeating cycle producing and enhancing inflammation (Gkogkolou and Böhm 2012).

As above explained, the problem of AGE production is strictly correlated with starting 'raw materials', including obviously foods and beverages (Van Nguyen 2006). The following compounds or extracts may inhibit in vivo AGE production, for evident therapeutic or inhibitory purposes (Edelstein and Brownlee 1992; Gkogkolou and Böhm 2012; Löbner et al. 2015; Navarro and Morales 2016; Rahbar and Figarola 2003; Totlani and Peterson 2005):

(f) Reducing agents such as ascorbate
(g) 'Nutraceuticals' (selenium yeast, riboflavin, etc.), and vitamins C (L-ascorbic acid) and E (α-tocopherol)
(h) Nucleophilic hydrazines (example: aminoguanidine)
(i) Specific 'AGE breakers' such as N-phenacylthiazolium (they can break protein crosslinks), or AGE scavengers such as pyridoxamine against Amadori products (via formation of a specific adduct). In general, the action against dicarbonyl compounds is observed. Other agents include creatine, hydroxytyrosol, cysteine, glutathione, and certain flavonoids.
(j) Enzymes against various Maillard reaction intermediates such as:

> (e.1) Oxoreductases (target: reducing sugars; obtained products: lactones). Unfortunately, the consequent production of hydrogen peroxide may be a problem (Petersen and Søe 2002; Søe and Petersen 2005).
> (e.2) Fructosamine oxidase: Target: Amadori products; desired reaction: oxidative deglycation (Troise et al. 2013).

With relation to some of these countermeasures, it can be reported in detail that (Bakris et al. 2004; Chen and Cerami 1993; Degenhardtet al. 2002; Edelstein and Brownlee 1992; Hirsch et al. 1992; Singh et al. 2001; Vasan et al. 2003; Voziyan and Hudson 2005):

(a) Pyridoxamine (Fig. 1.4) acts normally against obtained Amadori products with the aim of avoiding their AGE conversion.

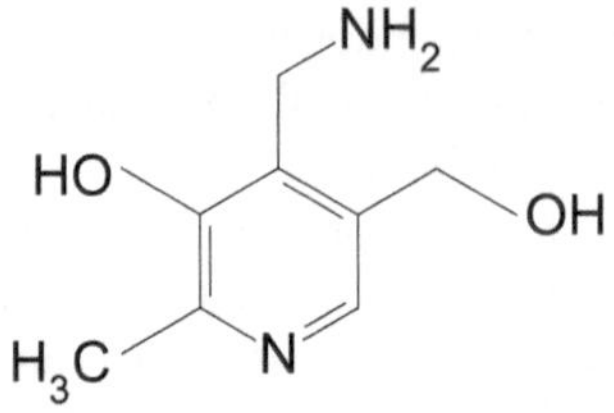

Fig. 1.4 Molecular structure of pyridoxamine, known as one of the vitamin B$_6$ forms. Pyridoxamine acts normally against obtained Amadori products with the aim of avoiding their AGE conversion. BKChem version 0.13.0, 2009 (http://bkchem.zirael.org/index.html) has been used for drawing this structure

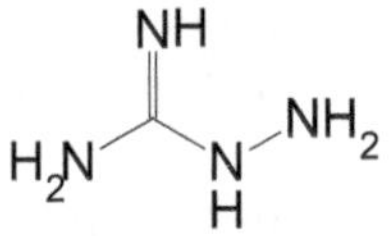

Fig. 1.5 Molecular structure of aminoguanidine. This nucleophilic hydrazine can act as a chemical trap for di-carbonyl molecules with the aim of avoiding AGE conversion of these compounds. It may also contrast crosslinking. Several results have been obtained when speaking of diabetes, nephropathy, retinopathy, and diabetic diseases. BKChem version 0.13.0, 2009 (http://bkchem.zirael.org/index.html) has been used for drawing this structure

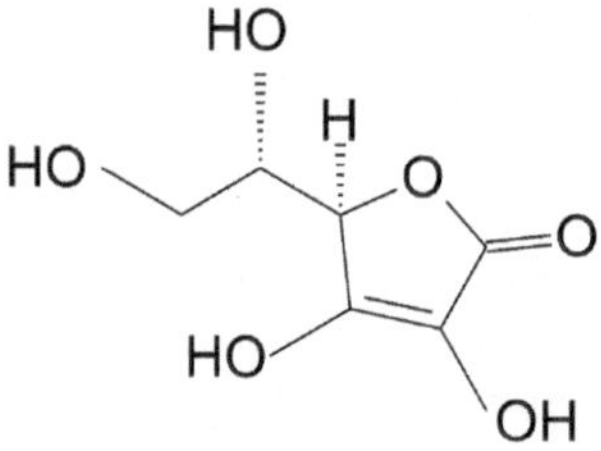

Fig. 1.6 Molecular structure of vitamin C (*L*-ascorbic acid). This nutraceutical (the heterogeneous group comprehends also vitamin E or α-tocopherol, selenium yeast, riboflavin, etc.) may interrupt the oxidative AGE conversion of di-carbonyl molecules, with results similar to the action of aminoguanidine. Vitamin C and E may also contrast some diabetic retinopathy, although these results have to be confirmed in human beings. BKChem version 0.13.0, 2009 (http://bkchem.zirael.org/index.html) has been used for drawing this structure

(b) Aminoguanidine (Fig. 1.5) can act as a chemical trap for di-carbonyl molecules with the aim of avoiding AGE conversion of these compounds. It may also act against crosslinking attitude. Several results have been obtained when speaking of diabetes, nephropathy, retinopathy, and diabetic diseases in general.

(c) Vitamin C (Fig. 1.6) may interrupt the oxidative conversion of di-carbonyl molecules to the final destination (AGE). Vitamins C and E may also contrast some diabetic retinopathy, although these results have to be confirmed in human beings.

On the other side, AGE can be remarkably produced by means of high processing temperatures for food productions (roasting, cooking, sterilisation,…); prolonged storage at high temperatures is also reported to have some influence (Lund and Ray 2017).

Finally, dietary restrictions and countermeasures may have some effect when speaking of limited AGE production and deposit into living tissues. Positive effects should include:

(a) The increase of average lifespan in humans
(b) The decrease of natural age dysfunctions
(c) The general diminution of dietary glycotoxins which generally determine the production of inflammatory agents with different negative effects including diabetes, renal failures, and skin ageing (Gkogkolou and Böhm 2012; Lyons et al. 1991; Peppa et al. 2003; Vlassara and Striker 2011; Yamagishi et al. 2007).

References

Ahmed N (2005) Advanced glycation endproducts-role in pathology of diabetic complications. Diab Res Clin Pract 67(1):3–21. https://doi.org/10.1016/j.diabres.2004.09.004

Ahmed MU, Brinkmann Frye E, Degenhardt TP, Thorpe SR, Baynes JW (1997) N-epsilon-(carboxyethyl)lysine, a product of the chemical modification of proteins by methylglyoxal, increases with age in human lens protein. Biochem J 324(2):565–570. https://doi.org/10.1042/bj3240565

ALjahdali N, Carbonero F (2017) Impact of Maillard reaction products on nutrition and health: Current knowledge and need to understand their fate in the human digestive system. Crit Rev Food Sci Nutr, 1–14. https://doi.org/10.1080/10408398.2017.1378865

Anjali Jain A (2004) Fighting obesity. BMJ 328:1327. https://doi.org/10.1136/bmj.328.7452.1327

Bakris GL, Bank AJ, Kass DA, Neutel JM, Preston RA, Oparil S (2004) Advanced glycation endproduct cross-link breakers. A novel approach to cardiovascular pathologies related to the aging process. Am J Hypertens 17(12):23S–30S. https://doi.org/10.1016/j.amjhyper.2004.08.022

Barbera M, Gurnari G (2017) Wastewater treatment and reuse in the food industry. Springer International Publishing, Cham

Baynes JW (2002) The Maillard hypothesis on aging: time to focus on DNA. Ann NY Acad Sci 959(1):360–367. https://doi.org/10.1111/j.1749-6632.2002.tb02107.x

Brunazzi G, Parisi S, Pereno A (2014) The importance of packaging design for the chemistry of food products. Springer International Publishing, Cham

Chen HJC, Cerami A (1993) Mechanism of inhibition of advanced glycosylation by aminoguanidine in vitro. J Carbohydr Chem 12(6):731–742. https://doi.org/10.1080/07328309308019003

Chen XM, Dai Y, Kitts DD (2016) Detection of Maillard reaction product [5-(5,6-Dihydro-4H-pyridin-3-ylidenemethyl)furan-2-yl]methanol (F3-A) in breads and demonstration of bioavailability in Caco-2 intestinal cells. J Agric Food Chem 64(47):9072–9077. https://doi.org/10.1021/acs.jafc.6b0436

Degenhardt TP, Alderson NL, Arrington DD, Beattie RJ, Basgen JM, Steffes MW, Thorpe SR, Baynes JW (2002) Pyridoxamine inhibits early renal disease and dyslipidemia in the streptozotocin-diabetic rat. Kidney Int 61(3):939–950. https://doi.org/10.1046/j.1523-1755.2002.00207.x

Delgado-Andrade C (2014) Maillard reaction products: some considerations on their health effects. Clin Chem Lab Med 52(1):53–60. https://doi.org/10.1515/cclm-2012-0823

Devcich DA, Pedersen IK, Petrie KJ (2007) You eat what you are: modern health worries and the acceptance of natural and synthetic additives in functional foods. Appet 48(3):333–337. https://doi.org/10.1016/j.appet.2006.09.014

Dyer DG, Dunn JA, Thorpe SR, Bailie KE, Lyons TJ, McCance DR, Baynes JW (1993) Accumulation of Maillard reaction products in skin collagen in diabetes and aging. J Clin Invest 91(6):2463–2469

Edelstein D, Brownlee M (1992) Mechanistic studies of advanced glycosylation end product inhibition by aminoguanidine. Diabetes 41(1):26–29. https://doi.org/10.2337/diab.41.1.26

Friedman M (2003) Chemistry, biochemistry, and safety of acrylamide. A review. J Agric Food Chem 51:4504–4526. https://doi.org/10.1021/jf030204+

Fu MX, Wells-Knecht KJ, Blackledge JA, Lyons TJ, Thorpe SR, Baynes JW (1994) Glycation, glycoxidation, and cross-linking of collagen by glucose: kinetics, mechanisms, and inhibition of late stages of the Maillard reaction. Diabetes 43(5):676–683. https://doi.org/10.2337/diab.43.5.676

Gkogkolou P, Böhm M (2012) Advanced glycation end products: key players in skin aging? Derm Endocrinol 4(3):259–270. https://doi.org/10.4161/derm.22028

Gurnari G (2015) Safety protocols in the food industry and emerging concerns. Springer International Publishing, Cham

Hayase F, Usui T, Nishiyama K, Sasaki S, Shirahashi Y, Tsuchiya N, Numata N, Watanabe H (2005) Chemistry and biological effects of melanoidins and glyceraldehyde-derived pyridinium as advanced glycation end products. Ann NY Acad Sci 1043(1):104–1010. https://doi.org/10.1196/annals.1333.013

Hellwig M, Bunzel D, Huch M, Franz CMAP, Kulling SE, Henle T (2015) Stability of individual Maillard reaction products in the presence of the human colonic microbiota. J Agric Food Chem 63(30):6723–6730. https://doi.org/10.1021/acs.jafc.5b01391

Henle T (2008) Maillard reaction of proteins and advanced glycation end products (AGEs) in food. In: Stadler RH, Lineback DR (eds) Process-induced food toxicants: occurrence, formation, mitigation, and health risks. Wiley, Hoboken. https://doi.org/10.1002/9780470430101.ch2g

Hirsch J, Petrakova E, Feather MS (1992) The reaction of some dicarbonyl sugars with aminoguanidine. Carbohydr Res 232(1):125–130. https://doi.org/10.1016/s0008-6215(00)90999-6

Hong H (2016) Modern industrial foods and their effects on the human body. Nat Med J 8, 4. Available https://www.naturalmedicinejournal.com/journal/2016-04/modern-industrial-foods-and-their-effects-human-body. Accessed 29 Oct 2018

Ilchmann A, Burgdorf S, Scheurer S, Waibler Z, Nagai R, Wellner A, Yamamoto Y, Yamamoto H, Henle T, Kurts C, Kalinke U, Vieths S, Toda M (2010) Glycation of a food allergen by the Maillard reaction enhances its T-cell immunogenicity: role of macrophage scavenger receptor class A type I and II. J Allergy Clin Immunol 125(1):175–183. https://doi.org/10.1016/j.jaci.2009.08.013

John WG, Lamb EJ (1993) The Maillard or browning reaction in diabetes. Eye (London) 7(Pt 2):230–237. https://doi.org/10.1038/eye.1993.55

Kirkland JL (2002) The biology of senescence: potential for prevention of disease. Clin Geriatr Med 18(3):383–405. https://doi.org/10.1016/s0749-0690(02)00023-x

Kokkinidou S (2013) Inhibition of Maillard reaction pathways and off-flavor development in UHT milk: structure reactivity of phenolic compounds. Dissertation, University of Minnesota, Minneapolis

Koschinsky T, He CJ, Mitsuhashi T, Bucala R, Liu C, Buenting C, Heitmann K, Vlassara H (1997) Orally absorbed reactive glycation products (glycotoxins): an environmental risk factor in diabetic nephropathy. Proc Natl Acad Sci USA 94(12):6474–6479. https://doi.org/10.1073/pnas.94.12.6474

Löbner J, Degen J, Henle T (2015) Creatine Is a scavenger for methylglyoxal under physiological conditions via formation of N-(4-Methyl-5-oxo-1-imidazolin-2-yl)sarcosine (MG-HCr). J Agric Food Chem 63(8):2249–2256. https://doi.org/10.1021/jf505998z

Lobo V, Patil A, Phatak A, Chandra N (2010) Free radicals, antioxidants and functional foods: impact on human health. Pharmacogn Rev 4(8):118–126. https://doi.org/10.4103/0973-7847.70902

Lund MN, Ray CA (2017) Control of Maillard reactions in foods: strategies and chemical mechanisms. J Agric Food Chem 65(23):4537–4552. https://doi.org/10.1021/acs.jafc.7b00882

Lyons TJ, Bailie KE, Dyer DG, Dunn JA, Baynes JW (1991) Decrease in skin collagen glycation with improved glycemic control in patients with insulin-dependent diabetes mellitus. J Clin Invest 87(6):1910–1915. https://doi.org/10.1172/JCI115216

Navarro M, Morales FJ (2016) In vitro investigation on the antiglycative and carbonyl trapping activities of hydroxytyrosol. Eur Food Res Technol 242(7):1101–1110. https://doi.org/10.1007/s00217-015-2614-8

Odetti P, Aragno I, Garibaldi S, Valentini S, Pronzato MA, Rolandi R (1998) Role of advanced glycation endproducts in aging collagen. A scanning force microscope study. Gerontol 44(4):187–191. https://doi.org/10.1159/000022008

Parisi S (2002) Profili evolutivi dei contenuti batterici e chimico-fisici in prodotti lattiero-caseari. Ind Aliment 41(412):295–306

Parisi S (2003) Evoluzione chimico-fisica e microbiologica nella conservazione di prodotti lattiero - caseari. Ind Aliment 42(423):249–259

Parisi S (2004) Alterazioni in imballaggi metallici termicamente processati. Gulotta Press, Palermo

Parisi S (2012) Food packaging and food alterations. The user-oriented approach. Smithers Rapra Technology Ltd., Shawbury

Parisi S (2013) Food industry and packaging materials. User-oriented Guidelines for Users. Smithers Rapra Technology Ltd., Shawbury

Parisi S, Luo W (2018) Chemistry of Maillard reactions in processed foods. Springer International Publishing, Cham

Parisi S, Delia S, Laganà P (2004) Il calcolo della data di scadenza degli alimenti: la funzione Shelf Life e la propagazione degli errori sperimentali. Ind Aliment 43(438):735–749

Parisi S, Laganà P, Delia S (2006) Curve di crescita dei miceti in diversi formaggi in atipiche condizioni di conservazione. Ind Aliment 45(458):532–538

Paul RG, Bailey AJ (1996) Glycation of collagen: the basis of its central role in the late complications of ageing and diabetes. Int J Biochem Cell Biol 28(12):1297–1310. https://doi.org/10.1016/S1357-2725(96)00079-9

Peppa M, Brem H, Ehrlich P, Zhang JG, Cai W, Li Z, Croitoru A, Thung S, Vlassara H (2003) Adverse effects of dietary glycotoxins on wound healing in genetically diabetic mice. Diabetes 52(11):2805–2813. https://doi.org/10.2337/diabetes.52.11.2805

Petersen L, Søe JB (2002) A process for the prevention and/or reduction of Maillard reaction in a foodstuff containing a protein, a peptide or an amino acid and a reducing sugar. World Intellectual Property Organization Patent WO2002039828A3, 23 Mar 2002

Popping B, Allred L, Bourdichon F, Brunner K, Diaz-Amigo C, Galan-Malo P, Lacorn M, North J, Parisi S, Rogers A, Sealy-Voyksner J, Thompson T, Yeung J (2018) Stakeholders' guidance document for consumer analytical devices with a focus on gluten and food allergens. J AOAC 101(1):1–5. https://doi.org/10.5740/jaoacint.17-0425

Rahbar S, Figarola JL (2003) Novel inhibitors of advanced glycation endproducts. Arch Biochem Biophys 419(1):63–79. https://doi.org/10.1016/j.abb.2003.08.00

Raj DSC, Choudhury D, Welbourne TC, Levi M (2000) Advanced glycation end products: a nephrologist's perspective. Am J Kidney Dis 35(3):365–380. https://doi.org/10.1016/S0272-6386(00)70189-2

Ramasamy R, Vannucci SJ, Yan SSD, Herold K Yan SF, Schmidt AM (2005) Advanced glycation end products and RAGE: a common thread in aging, diabetes, neurodegeneration, and inflammation. Glycobiol 15(7):16R–28R. https://doi.org/10.1093/glycob/cwi053

Sanders TAB (1999) Food production and food safety. BMJ 318(7199):1689–1693. https://doi.org/10.1136/bmj.318.7199.1689

Schleicher E, Wagner E, Nerlich A (1997) Increased accumulation of glycoxidation product carboxymethyllysine in human tissues in diabetes and aging. J Clin Invest 99(3):457–468. https://doi.org/10.1172/jci119180

Scott E (2003) Food safety and foodborne disease in 21st century homes. Can J Infect Dis 14(5):277–280. https://doi.org/10.1155/2003/363984

Šebeková K, Saavedra G, Zumpe C, Somoza V, Klenovicsová K, Birlouez-Aragon I (2008) Plasma concentration and urinary excretion of $N\varepsilon$-(Carboxymethyl)lysine in breast milk- and formula-fed infants. Ann NY Acad Sci 1126(1):177–180. https://doi.org/10.1196/annals.1433.049

Seiquer I, Rubio LA, Peinado MJ, Delgado-Andrade C, Navarro MP (2014) Maillard reaction products modulate gut microbiota composition in adolescents. Mol Nutr Food Res 58(7):1552–1560. https://doi.org/10.1002/mnfr.201300847

Simpson BK (2012) Food biochemistry and food processing, 2nd edn. Wiley-Blackwell, Ames

Singh R, Barden A, Mori T, Beilin L (2001) Advanced glycation end-products: a review. Diabetologia 44(2):129–146. https://doi.org/10.1007/s001250051591

Singla RK, Dubey AK, Ameen SM, Montalto S, Parisi S (2018) Analytical methods for the assessment of Maillard reactions in foods. Springer International Publishing, Cham

Søe JB, Petersen LW (2005) Methods of reducing or preventing Maillard reactions in potato with hexose oxidase. US Patent 6,872,412 B2, 29 Mar 2005

Tessier FJ, Jacolot P, Niquet-Leridon C (2014) Research commentaries for the members of the international Maillard reaction society: open questions around the carboxymethyllysine. IMARS Highlights 9(3):14–20

Thorpe SR, Baynes JW (2003) Maillard reaction products in tissue proteins: new products and new perspectives. Amino Acids 25(3–4):275–281. https://doi.org/10.1007/s00726-003-0017-9

Toda M, Heilmann M, Ilchmann A, Vieths S (2014) The Maillard reaction and food allergies: is there a link? Clin Chem Lab Med 52(1):61–67. https://doi.org/10.1515/cclm-2012-0830

Totlani VM, Peterson DG (2005) Reactivity of epicatechin in aqueous glycine and glucose Maillard reaction models: Quenching of C2, C3, and C4 sugar fragments. J Agric Food Chem 53(10):4130–4135. https://doi.org/10.1021/jf050044x

Traverso N, Menini S, Maineri EP, Patriarca S, Odetti P, Cottalasso D, Marinari UM, Pronzato MA (2004) Malondialdehyde, a lipoperoxidation-derived aldehyde, can bring about secondary oxidative damage to proteins. J Gerontol Series A: Biol Sci Med Sci 59(9):B890–B895. https://doi.org/10.1093/gerona/59.9.b890

Troise AD, Dathan NA, Fiore A, Roviello G, Di Fiore A, Caira S, Cuollo M, De Simone G, Fogliano V, Monti SM (2013) Faox enzymes inhibited Maillard reaction development during storage both in protein glucose model system and low lactose UHT milk. Amino Acids 46(2):279–288. https://doi.org/10.1007/s00726-013-1497-x

Trowell H (1972) Ischemic heart disease and dietary fiber. Am J Clin Nutr 25(9):926–932. https://doi.org/10.1093/ajcn/25.9.926

Urribarri J, Cai W, Sandu O, Peppa M, Goldberg T, Vlassara H (2005) Diet-derived advanced glycation end products are major contributors to the body's AGE pool and induce inflammation in healthy subjects. Ann NY Acad Sci 1043(1):461–466. https://doi.org/10.1196/annals.1333.052

Van Nguyen C (2006) Toxicity of the AGEs generated from the Maillard reaction: on the relationship of food-AGEs and biological-AGEs. Mol Nutr Food Res 50(12):1140–1149. https://doi.org/10.1002/mnfr.200600144

Vasan S, Foiles P, Founds H (2003) Therapeutic potential of breakers of advanced glycation end product- protein crosslinks. Arch Biochem Biophys 419(1):89–96. https://doi.org/10.1016/j.abb.2003.08.016

Vlassara H (2005) Advanced glycation in health and disease: role of the modern environment. Ann NY Acad Sci 1043(1):452–460. https://doi.org/10.1196/annals.1333.051

Vlassara H, Striker GE (2011) AGE restriction in diabetes mellitus: a paradigm shift. Nat Rev Endocrinol 7(9):526–539. https://doi.org/10.1038/nrendo.2011.74

Vlassara H, Cai W, Crandall J, Goldberg T, Oberstein R, Dardaine V, Peppa M, Rayfield EJ (2002) Nonlinear partial differential equations and applications: Inflammatory mediators are induced by dietary glycotoxins, a major risk factor for diabetic angiopathy. Proc Natl Acad Sci USA 99(24):15596–15601. https://doi.org/10.1073/pnas.242407999

Voziyan PA, Hudson BG (2005) Pyridoxamine: the many virtues of a Maillard reaction inhibitor. Ann NY Acad Sci 1043(1):807–816. https://doi.org/10.1196/annals.1333.093

WHO/FAO (2003) Diet, nutrition and the prevention of chronic diseases. Report of a joint WHO/FAO expert consultation. WHO technical report series No 916. World Health Organization (WHO), Geneva, and Food and Agriculture Organization of the United Nations (FAO), Rome

Yamagishi S, Ueda S, Okuda S (2007) Food-derived advanced glycation end products (AGEs): a novel therapeutic target for various disorders. Curr Pharm Des 13(27):2832–2836. https://doi.org/10.2174/138161207781757051

Zhang Q, Ames JM, Smith RD, Baynes JW, Metz TO (2008) A perspective on the Maillard reaction and the analysis of protein glycation by mass spectrometry: probing the pathogenesis of chronic disease. J Proteom Res 8(2):754–769. https://doi.org/10.1021/pr800858h

Chapter 2
The Amount of Carbohydrates in the Modern Diet and the Influence of Food Taxes for Public Health Purposes

Abstract This chapter discusses the importance of carbohydrates in inexpensive and good-tasting foods. The behaviour of food consumers has been always influenced by material and immaterial features concerning food and beverage products. Several of these features are related to the marketing strategy, advertising messages, and so on. However, a good and relevant part of motivating reasons concerns the food products 'as it is'. This reflection involves chemistry, microbiology, and technology of foods and beverages. For these reasons, the industry tends to produce tasty and commonly accepted food products. In relation to sweet products, the synergic effect of fat and sugars at the hedonistic level should be considered. Food preference depends often on sweet tastes, but this attribution can be ascribed to the lipid content in certain foods such as cheeses without high sugar contents. Moreover, the amount of sugars in the human diet is not a direct cause for obesity. Consequently, different strategies have been proposed when speaking of the fight against malnutrition and obesity, including nutritional education, health training, nutritional labellings, marketing campaigns, use of different sugars and fat matters, and the so-called food tax policies. These strategies are considered in detail, especially when speaking of tax policies and sweet surrogates.

Keywords Acesulfame potassium · Added sugar · Food tax · Marketing · Obesity · Sucrose · Sweetening power

Abbreviations

FAO	Food and Agriculture Organization of the United Nations
DIETFITS	Diet Intervention Examining the Factors Interacting with Treatment Success
SP	Sweetening power
USDA	US Department of Agriculture

© The Author(s), under exclusive license to Springer Nature Switzerland AG 2019   15
S. D. Sharma and M. Barone, *Dietary Patterns, Food Chemistry and Human Health*,
Chemistry of Foods, https://doi.org/10.1007/978-3-030-14654-2_2

2.1   Good-Tasting Food Products Today

The behaviour of food consumers—both in the past and at present—has been always influenced by material and immaterial features concerning food and beverage products. Several of these features are related to the marketing strategy, advertising messages, and so on, especially in the modern era of social networks (Cairns et al. 2009; Institute of Medicine et al. 2006). However, a good and relevant part of motivating reasons concerns the food products 'as it is'. This reflection involves chemistry, microbiology, and technology of foods and beverages. For these reasons, the industry tends to produce tasty and commonly accepted food products, also named 'comfort foods' (Wansink et al. 2003).

Material features have to be interpreted by consumers, and this interpretation is challenging enough. On the one side, the food should be judged for its intrinsic and visually appreciable content; on the other side, physiological and psychological reasons could influence the consumer in a number of different ways. Basically, the preference for one or another food can be influenced without the conscious control of the human consumer, but the knowledge of certain food attribution can be critical. In other terms, 'normal' consumers—the average people able to buy, make choices, and judge foods and beverages with a direct effect on the market (Brunazzi et al. 2014; Parisi 2012, 2013)—can receive more pleasure from certain foods, also named 'comfort foods' (Wansink et al. 2003). The psychological influence cannot be excluded, especially when speaking of infant and non-adult consumers (Bradley 1972; Bradley and Mistretta 1975; Le Magnen 1986; Stewart et al. 1984; Wise 1988). Also, psychological preferences have to be carefully considered when speaking of insufficient dietary variety as perceived by the food consumer (Rolls 1986, 1999; Rolls and McDermott 1991), desire of more palatable foods without conscious participation (Pliner and Melo 1997; Wansink 1994; Wansink et al. 2003), peculiar stress conditions enhancing the consumption of fat, sweet, or general tasty foods (McCann et al. 1990; Oliver and Wardle 1999), dependency from the assumption of peculiar compounds (Le Magnen 1986; McCloy and McCloy 1979; Mistretta 1981), and early food consumption habits or affective motivations in the first years at least (Barthel 1989; Wansink 2002; Wansink et al. 2002).

On the other hand, psychological motivations have to be also considered with attention (Arnow et al. 1995; Birch et al. 1989; Galef 1991; Tuomisto et al. 1998; Wansink et al. 2003). Three general factors at least may influence the behaviour of 'conscious' normal consumers (Birch et al. 1980; Booth 1985; DeCastro and DeCastro 1989; Wansink 2003; Wansink et al. 2002, 2003):

(a) The social environment
(b) The identification of the food consumer in the context of social relationships
(c) The conditioned answer of the food consumer in response to adequate external stimuli.

In relation to social environments, human and animal consumers are always influenced from the behaviour of their social context, with the consequent 'imitation' relationships between the consumer A and other subjects B, C, etc. As a result, should a main part of consumers (with the exclusion of A) make a certain purchase, or enjoy a peculiar product, the consumer A would probably be influenced—consciously or not—in this way (Frijda et al. 1989; Houston 1979; Macht 1999; Slochower 1983). The modification of food preferences probably causes the subsequent information for food industries and related non-food segments (this list may not be exhaustive):

(1) The food 'as it is' is favourably accepted by the consumer A in a social environment preferring the food, and this acceptance may probably be increased in a broader environment where consumers B, C, etc., are more abundant and strictly connected.
(2) The food is not favourably accepted by the consumer A, while the other consumers preferring the food are apparently disconnected from consumer A (antisocial behaviour limits or inhibits the social influence).
(3) The food is not favourably accepted by the consumer A. However, the other consumers do not prefer the food, and they are apparently disconnected from consumer A (antisocial behaviour does not limit or inhibit the social influence).
(4) The food is favourably accepted by the consumer A, but the other consumers preferring the food are apparently disconnected from consumer A (antisocial behaviour does not limit or inhibit the social influence).
(5) The food is not favourably accepted by the consumer A because of the non-general acceptance (or refusal) by other consumers. In this situation, the influence of the social context on the consumer A is extremely relevant.

The identification of the food consumer, including children, in the context of social relationships (at home, school, workplace, etc.) is essentially linked with psychological features such as the relation: expectation/gift or satisfaction. As a result, social experiences related to peculiar events in the early life of food consumers and in the adult life can create easy associations between a more or less palatable food and the consumer (Cairns et al. 2009; Cowart and Beauchamp 1986; Drewnowski and Greenwood 1983; Galef 1991; Harris 2008; Wansink 1994; Weingarten and Kulikovsky 1989). Actually, these experiences can be both negative and positive and explain a certain variety of situations; anyway, these events have to be also correlated with material and intrinsic food properties (salty taste, fluidity, colorimetric appearance, etc.). Interestingly, the refusal of certain foods and/or beverages in the childhood is reported as 'neophobic answer', suggesting a distinct phobia of children for several food types (Harris 2008).

In specific relation to 'comfort foods', it has been reported that these products are favourably accepted by consumers depending on age; in addition, the accepted subdivision between snack-related products and meal-related foods seems to be correlated with age. In other words:

(1) The trend in favour of more palatable foods appears increasing when speaking of youngers, in general.
(2) Also, snack foods (pronounced saltiness or sweetness) seem particularly desired by youngers, while older consumers are not so attracted.
(3) The preference versus meal-related products appears clear enough when speaking of older consumers.

(4) Consequently, snack foods and similar products (consumption out of the normal
     household meal: French fries, cookies, candies, etc.) are generally preferred by
     youngers, while older consumers—often accustomed to eating at home—prefer
     to buy and evaluate favourably meal-related products (salads, vegetable soups,
     Italian pasta, pizza, etc.).

Interestingly, these results (Wansink et al. 2003) seem to be influenced by first
consumption experiences because children accustomed to eat snack foods appear
willing to replicate this behaviour in the adult age (and probably not accustomed to
comprehend the complexity of different tastes). For the same—or similar reasons—it
has been also reported that females are more willing to eat snack foods and similar
products, while males appear to appreciate meals (Cowart and Beauchamp 1986;
Desor et al. 1975; Food Marketing Institute and Better Homes and Gardens Magazine 1988; Mennella and Beauchamp 1996; Wansink 2002; Wansink and Sudman
2002). Because of these results, the relationship between palatable food products and
consumers (in terms of categories) may be precious when speaking of public health
campaigns against obesity, malnutrition, and eating disorders in general (Arnow et al.
1992, 1995; Blundell and Hill 1993; Kaplan and Kaplan 1957). Also, these products
are generally defined 'junk foods' for several reasons; however, the general appre-
ciation for these products in Western-style society appears stronger than detailed
guidelines diffused by means of normal media and also social networks.

Another important feature of the problem is represented by economic concerns.
In fact, palatable foods—including 'junk foods'—are often associated with cheaper
choices, while expensive products are associated with higher money availability on
the one side and high-value tastes on the other side (e.g. historical recipes, regional
products). The possibility of ethical concerns and behavioural habits related to food
eating can further complicate the problem (Drewnowski 1995, 1997; Logue 1986;
Perissé et al. 1969; Popkin 1994; Sanjur 1982). For these reasons, the recent policy of
food taxes in certain countries has been used. Section 2.4 is dedicated to this aspect
of public health measures.

In addition, it should be considered that the progressive augments of wellness
in industrialised countries have been defined one of the co-causes for the increasing trend of meat-, milk-, and sugar-based diets (Popkin 1994; Popkin et al. 1995)
in contrast with grain-based habits. In these conditions, it has been reported that
modern diets give 30–40% of the total caloric content from lipids, while complex
carbohydrates and dietary fibres decrease if compared with fat matter and sugars
(Drewnowski 1995; Popkin 1994).

2.2   Carbohydrates in Foods. The Sweet Choice
for Good-Tasting Products

By the chemical and nutritional viewpoints, it could be assumed that each dispropor-
tion between nutrients in the human diet is the cause of health effects. This simple

reflection highlights the role of nutrients, food by food, in the broad ambit of a balanced diet (day by day), instead of examining only peculiar food products. On the other hand, the role of certain foods is obvious when considering eating disorders, if these products are used frequently. In general, foods with relevant sugar and/or fat contents are extremely accepted by the normal consumer, while the combination of high sugar and fat amounts enhances 'selling hopes' (Drewnowski 1987; Drewnowski and Greenwood 1983). It should be also noted that the only presence of high sugar amounts does not appear to be the direct cause for human obesity, in spite of common opinions (Anderson 1995; Drewnowski 1997; Gibney et al. 1995). A recent research—the 'Diet Intervention Examining the Factors Interacting with Treatment Success' (DIETFITS)—has confirmed that genetic factors are not responsible for different answers—and the result—when speaking of low-fat or low-carbohydrate diets (Stanton et al. 2017).

With reference to several food categories with distinct sweet tastes, some researches based on sensorial studied carried out by panellists have demonstrated that hedonistic behaviour in consumers seems mainly influenced by fat and sweet tastes at the same time. In other words, the perceived sweetness seems more evident when sugars are present in—or added to—food formulation with a remarkable fat amount. Should sugar amount be about 8%, this effect would be confirmed when the ratio of fat/sugar reaches at least 2.5. This effect is particularly evident in certain cheeses and milk-based products (Drewnowski and Holden-Wiltse 1992; Drewnowski and Greenwood 1983; Drewnowski et al. 1985). Consequently, the first 'lesson' to be learned when speaking of sweet foods should concern the synergic effect of fat and sugar at the hedonistic level. The strategy is generally valuable in the sector of cheeses, where products such as mozzarella cheeses and specially processed cheeses exhibit a distinctly sweet taste with the corresponding augment of milk fat (usually exceeding 15%). On the other hand, it should be noted that these products contain 2% or less of lactose (and carbohydrates, in general), with a virtual absence in processed cheeses. It may be assumed that food preference for these products depends mainly on sweet tastes, but this attribution is only ascribed to the lipid content (Barbieri et al. 2014; Parisi 2002, 2003). Anyway, the synergic taste effect is also evident when speaking of products for children such as jelly sandwiches, cookies, and many candies (Drewnowski 1997; Kern et al. 1993).

On the other side, the influence of endogenous opioid peptides (such as butorphanol and naloxone) on sensorial preferences for fat- and sugar-containing products has been demonstrated in humans and in rats. As a result, it has been suggested that the behaviour of bulimic and anorectic patients may be dependent both on the ratio between lipids and carbohydrates (simple sugars) on the one hand, and on stimuli by means of these opioid peptides as correlated with these eating disorders (Drewnowski and Greenwood 1983; Drewnowki et al. 1985, 1987a, b, 1992). In detail, naloxone, in particular, seems a good mean when speaking of reduction of sugar/fat foods, although the reduction appears different when speaking of different patients.

2.3   Added Sugars, Foods, and Obesity. Public Health Countermeasures

As mentioned above(Sect. 2.2), the amount of sugars in the human diet is not a direct cause for obesity (Anderson 1995; Drewnowski 1997; Gibney et al. 1995).

At present, the following strategies should be mentioned when speaking of fight against malnutrition and obesity, in terms of public health policy, although some attempt could be considered in the ambit of industrial marketing and technological research (Contento et al. 1995; Drewnowski 1997; Gortmaker et al. 2011; Swinburn et al. 2011; USDA 1995; Vos et al. 2010):

(a) Nutritional education and training (at schools, communities, hospitals, etc.) focusing on the need for more nutritional foods instead of high-fat, high-salt, sweetened foods, and sugar-sweetened drinks
(b) Health training with the aim of favouring physical activity and reduction of television viewing
(c) Marketing campaigns for the promotion of high-fibre and 'healthy' products
(d) The so-called food tax policies aiming at the reduction of high-fat, high-salt, and sweetened products by means of forced price augments
(e) New nutritional labelling strategies such as the 'traffic light' models (Sacks et al. 2010)
(f) The use of different sugars and fat matters in the production of palatable foods
(g) Drastic surgery actions such as 'gastric banding'.

In relation to these strategies, it should be considered that obtained results can vary enough worldwide, country by country. The national action is prevailing, although several international agencies are involved in this ambit (Gortmaker et al. 2011). A reliable analysis is really challenging, and detailed researches have been reported to focus both on health effects and cost savings (because a hospitalised patient represents a remarkable cost). However, fight against obesity and other eating disorders needs more efforts at present.

An interesting model against the increasing trend of certain foods and beverages has been carried out in several countries in recent years, in relation to possible fiscal actions, similarly to other strategies against excessive tobacco use (Colchero et al. 2016; Jha et al. 2006; Levy et al. 2010). It should be noted that similar actions are not completely shared between different countries at present because of possible effects on food and beverage markets (Horton 2018).

2.4   Food Taxation Against Added Sugars. Is This the Good Strategy?

The winning strategy against excessive tobacco consumption has been based on a massive advertising campaign with the aim of showing an unhealthy effect on smokers and non-smokers. Also, an effective fiscal-pricing action on tobacco products has been carried out in recent years with interesting results. On the other hand, there are not specific and reliable evidences that similar actions can reduce effects of eating disorders, if fiscal policies and/or subsidies (for the consumption of 'healthy foods') concern 'added sugars' foods and beverages (Colchero et al. 2016; Powell and Chaloupka 2009). The declared aim is to limit the conscious consumption of these foods with a measurable effect on the number of patients with eating disorders. In theory, the action should have some effect on the economic level in controlled environments (where the subject is strictly controlled). In practice, some limitations have been reported when speaking of:

(a) Choice of food or beverage categories for food taxation or subsidies (depending also on the estimated low or high price per item, and the related abundance on the market)
(b) Uncontrolled conditions of the study (in other words, uncontrolled environment)
(c) The magnitude of tax action if compared with food prices
(d) Correct information to the consumer (e.g. justification for tax actions as a measure against obesity should be explicitly promoted as a means to obtain funds against disorders)
(e) Dimensions of the market into a country
(f) Economic availability of common consumers.

In summary, it appears that small taxation does not easily reduce the consumption of added sugar foods and beverages (a similar campaign can be carried out against high-fat foods, but the results might be similar). On the opposite hand, these measures could have a general impact on the whole population in a country on condition that (Powell and Chaloupka 2009):

(1) Other measures are taken at the same time (Sect. 2.3).
(2) The tax effect is numerically high in relation to the estimated item price and the income of consumers.

Anyway, the multifaceted approach appears the best choice, and it should be able to have measurable effects on the pre-adult population if performed broadly and during a long temporal period. In this ambit, the choice of certain artificial sweeteners could give some help.

2.5 Carbohydrates and Palatable Foods. Chemical Countermeasures Against Obesity and Hypernutrition

From the chemical viewpoint, sweetened foods and palatable products show a high amount of one or more of the following sugars (Anderson 1995; Barbieri et al. 2014; Bray and Popkin 1998; Sclafani 1991):

(a) Sucrose

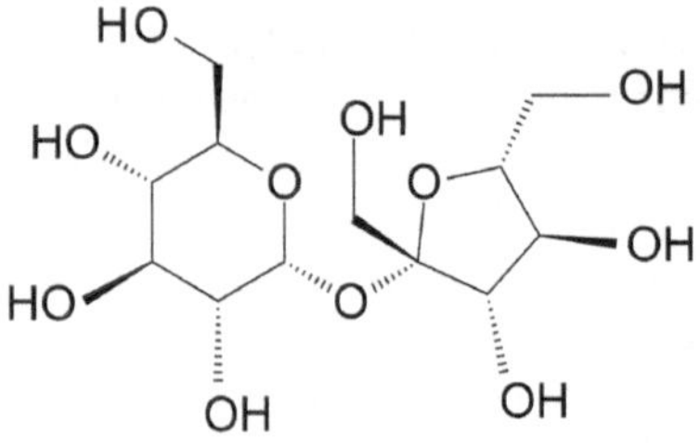

Fig. 2.1 Molecular structure of sucrose. BKChem version 0.13.0, 2009 (http://bkchem.zirael.org/index.html) has been used for drawing this structure

(b) Glucose
(c) Fructose (this sugar may be present or added 'as it is' or as high-fructose corn syrup). In particular, 'free fructose' is the mono-saccharide (non-bound) form only, while the expression 'total fructose' concerns the sum of free and non-free (bound) fructose molecules
(d) Maltose
(e) Galactose.

Interestingly, the definition 'added sugar' concerns all sugars—mono- or disaccharides—added to foods: the above-mentioned sucrose and high-fructose corn syrup are good examples, but the list can comprehend many syrups, honey, etc. (Baglio 2017). As a result, the problem of 'added sugars' appears linked to related definitions: 'naturally occurring sugars' exclude added sugars, and 'total sugars' concern the sum of naturally occurring and added foods. One of these sugars, sucrose, is shown in Fig. 2.1.

The sweetness of modern foods and beverages is mainly ascribed to the addition of syrups—such as sweet corn-based syrups—to food productions. These products, after treatment with an enzymatic glucose isomerase, can hydrolyse sucrose—with sweetening power (SP) defined '100—with the production of a 1:1 mixture between glucose[1] and fructose[2] (Krause and Mahan 1984). Consequently, 40% of the total amounts of sweeteners today are represented by these cheap hydrolysed masses (Bray et al. 2004; Putnam and Allshouse 1970). For this reason, it has been reported that a certain part of obesity-related problems for patients using sweetened products could be correlated with the augment of free fructose: it appears that the caloric abundance of high-fructose corn syrups if compared with total caloric sweeteners has increased in sweetened foods up to 19.7% in 2000 in relation to the year 1985, with reference to the USA (Putnam and Allshouse 1970). However, more research is still needed in this ambit before taking reliable conclusions.

The substitution or replacement of these sugars with artificial and non-nutritive sweeteners, such as aspartame, acesulfame potassium, saccharin, and stevia extracts from *Stevia rebaudiana* Bertoni, is the 'chemical' answer of the food industry to

---

[1] SP: 74 if compared with sucrose.
[2] SP: 174 (minimum value).

the problem of palatability (Anton et al. 2010; Cramer and Ikan 1987; Goyal et al. 2010; Lemus-Mondaca et al. 2012; Phillips 1987; Sundaresan 2018). On the other hand, it has been demonstrated in rats that the consumption of sweetened foods with saccharin may cause the increase of body-weight values because of the augment of adipose tissue, similar to the consumption of sweetened foods with caloric sugars (Swithers and Davidson 2008). This situation may explain partially the explosion of obesity disorders in the industrialised world in spite of the increased consumption of 'healthy' and non-caloric foods, although the subdivision of patients in different social or ethnic groups has to be considered (Flegal et al. 2002; Freedman et al. 2001, 2004). Actually, it can be inferred that the psychological justification of health foods has continuously increased the consumption of sweetened foods per capita, with the consequent increase of global consumption (Swithers and Davidson 2008). It has also been reported that aspartame could be more effective than sucrose—a naturally occurring sugar—or stevia extracts, based on hedonic behaviours and evaluations. Consequently, these sweeteners might be effective as 'chemical' caloric reducing agents, while extract stevioside could be not good enough if compared with nutritive sucrose (Anton et al. 2010). Other hypocaloric molecules, sugar alcohols, can be used in the food ambit, including (Barbieri et al. 2014):

(1) Xylitol or (2R,3R,4S)-pentane-1,2,3,4,5-pentol, obtained from xylose by hydrogenation
(2) Erythritol or (2R,3S)-butane-1,2,3,4-tetraol), extracted from algae and fruits
(3) Maltitol or 4-O-α-D-glucopyranosyl-D-glucitol. It is obtained from maltose by hydrogenation.

The interest for sugars in animal and human food is growing in recent years, and the same thing can be affirmed when speaking of reliable analytical procedures able to detect them in foods (Szpylka et al. 2018); for this reason, the research is continually evolving in this heterogeneous ambit.

A good example representing the class of artificial and non-nutritive sweeteners is acesulfame potassium (Fig. 2.2). It is one of the best artificial sweeteners when speaking of general use in the industry of cosmetics, foods, pharma products, etc. Also, it is widely used in different formulations such as toothpaste preparations (Kloesel 2000; Schmidt et al. 2000). It is commonly accepted in many industrialised countries (European Union, USA, etc.), with a sweetening power at least equal to 180 times if compared with sucrose, and a synergistic flavouring system (the enhancement of natural odours is widely recognised). In addition, taste synergistic actions have been demonstrated if acesulfame potassium is used with aspartame or sodium cyclamate, while the use of sodium saccharine can decrease the sweetening effect (Rowe et al. 2006).

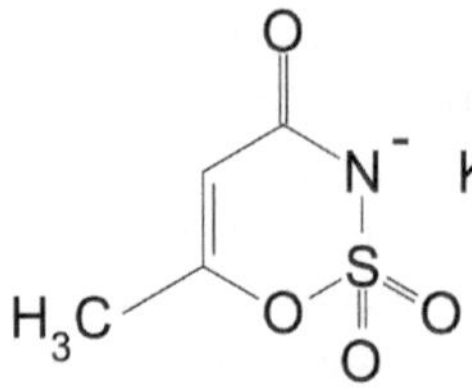

Fig. 2.2 Molecular structure of acesulfame potassium, also named E950 or 6-methyl-3,4-dihydro-1,2,3-oxathiazine-4(3H)-one 2,2-dioxide potassium salt. It is one of the best artificial sweeteners when speaking of general use in the industry of cosmetics, foods, pharma products, etc. It is commonly accepted in many industrialised countries with a sweetening power at least equal to 180 times if compared with sucrose, and a synergistic flavouring system. BKChem version 0.13.0, 2009 (http://bkchem.zirael.org/index.html) has been used for drawing this structure

References

Anderson CH (1995) Sugars, sweetness, and food intake. Am J Clin Nutr 62(1):195S–201S. https://doi.org/10.1093/ajcn/62.1.195s

Anton SD, Martin CK, Han H, Coulon S, Cefalu WT, Geiselman P, Williamson DA (2010) Effects of stevia, aspartame, and sucrose on food intake, satiety, and postprandial glucose and insulin levels. Appetite 55(1):37–43. https://doi.org/10.1016/j.appet.2010.03.009

Arnow B, Kenardy J, Agras WS (1992) Binge eating among the obese: a descriptive study. J Behav Med 15(2):155–170. https://doi.org/10.1007/bf00848323

Arnow B, Kenardy J, Agras WS (1995) The emotional eating scale: the development of a measure to assess coping with negative affect by eating. Int J Eat Disord 18(1):79–90. https://doi.org/10.1002/1098-108x(199507)18:1%3c79:aid-eat2260180109%3e3.0.co;2-v

Baglio E (2017) Chemistry and technology of honey production. Springer International Publishing, Cham

Barbieri G, Barone C, Bhagat A, Caruso G, Conley ZR, Parisi S (2014) Sweet compounds in foods: sugar alcohols. The influence of chemistry on new foods and traditional products. Springer International Publishing, Cham, pp 51–59

Barthel D (1989) Modernism and marketing: the chocolate box revisited. Theor Cult Soc 6(3):429–438. https://doi.org/10.1177/026327689006003004

Birch LL, Zimmerman SI, Hind H (1980) The influence of social-affective context on the formation of children's food preferences. Child Dev 51(3):856–861. https://doi.org/10.2307/1129474

Birch LL, McPhee L, Sullivan S, Johnson S (1989) Conditioned meal initiation in young children. Appetite 13(2):105–113. https://doi.org/10.1016/0195-6663(89)90108-6

Blundell JE, Hill AJ (1993) Binge eating: psychobiological mechanisms. In: Fairburn CG, Wilson GT (eds) Binge eating. Nature, assessment, and treatment. Guilford, London, pp 206–224

Booth DA (1985) Food-conditioned eating preferences and aversions with interceptive elements: conditioned appetites and satieties. Ann NY Acad Sci 443(1):22–41. https://doi.org/10.1111/j.1749-6632.1985.tb27061.x

Bradley RM (1972) Development of the taste bud and gustatory papillae in human fetuses. In: Bosma JF (ed) Third symposium on oral sensation and perception: the mouth of the infant. Charles C Thomas, Springfield, pp 139–142

Bradley RM, Mistretta C (1975) Fetal sensory receptors. Physiol Rev 55(3):182–352. https://doi.org/10.1152/physrev.1975.55.3.352

Bray GA, Popkin BM (1998) Dietary fat intake does affect obesity! Am J Clin Nutr 68(6):1157–1173. https://doi.org/10.1093/ajcn/68.6.1157

Bray GA, Nielsen SJ, Popkin BM (2004) Consumption of high-fructose corn syrup in beverages may play a role in the epidemic of obesity. Am J Clin Nutr 79(4):537–543. https://doi.org/10.1093/ajcn/79.4.537

Brunazzi G, Parisi S, Pereno A (2014) The importance of packaging design for the chemistry of food products. Springer International Publishing, Cham

Cairns G, Angus K, Hastings G (2009) The extent, nature and effects of food promotion to children: a review of the evidence to December 2008. World Health Organization, Geneva. Available https://www.who.int/dietphysicalactivity/publications/marketing_evidence_2009/en/. Accessed 30th Oct 2018

Colchero MA, Popkin BM, Rivera JA, Wen S (2016) Beverage purchases from stores in Mexico under the excise tax on sugar sweetened beverages: observational study. BMJ 352:h6704. https://doi.org/10.1136/bmj.h6704

Contento I, Balch GI, Bronner YL, Lytle LA, Maloney SK, Olson CM, Swadener SS (1995) The effectiveness of nutrition education and implications for nutrition education policy, programs, and research: a review of research. J Nutr Educ 27(6):277–418

Cowart BJ, Beauchamp GK (1986) The importance of sensory context in young children's acceptance of salty tastes. Child Dev 57(4):1034–1039. https://doi.org/10.1111/j.1467-8624.1986.tb00264.x

Cramer B, Ikan R (1987) Progress in the chemistry and properties of rebaudiosides. In: Grenby TH (ed) Developments in sweeteners. Elsevier, New York, pp 45–48

DeCastro JM, DeCastro E (1989) Spontaneous meal patterns of humans: influence of the presence of other people. Am J Clin Nutr 50(2):237–247. https://doi.org/10.1093/ajcn/50.2.237

Desor JA, Greene LS, Maller O (1975) Preference for sweet and salty in 9- and 15-year old and adult humans. Science 190(4215):686–687. https://doi.org/10.1126/science.1188365

Drewnowski A (1987) Fats and food texture: sensory and hedonic evaluations. In: Moskowitz HR (ed) Food texture. Marcel Dekker, New York, pp 217–250

Drewnowski A (1995) Energy intake and sensory properties of food. Am J Clin Nutr 62(5):1081S–1085S. https://doi.org/10.1093/ajcn/62.5.1081s

Drewnowski A (1997) Taste preferences and food intake. Annu Rev Nutr 17(1):237–253. https://doi.org/10.1146/annurev.nutr.17.1.237

Drewnowski A, Greenwood MR (1983) Cream and sugar: human preferences for high-fat foods. Physiol Behav 30(4):629–633. https://doi.org/10.1016/0031-9384(83)90232-9

Drewnowski A, Holden-Wiltse J (1992) Taste responses and food preferences in obese women: effects of weight cycling. Int J Obes 16:639–648

Drewnowski A, Brunzell JD, Sande K, Iverius PH, Greenwood MRC (1985) Sweet tooth reconsidered: taste responsiveness in human obesity. Physiol Behav 35(4):617–622. https://doi.org/10.1016/0031-9384(85)90150-7

Drewnowski A, Halmi KA, Pierce B, Gibbs J, Smith GP (1987a) Taste and eating disorders. Am J Clin Nutr 46(3):442–450. https://doi.org/10.1093/ajcn/46.3.442

Drewnowski A, Bellisle F, Aimez P, Remy B (1987b) Taste and bulimia. Physiol Behav 41(6):621–626. https://doi.org/10.1016/0031-9384(87)90320-9

Drewnowski A, Krahn D, Demitrack M, Nairn K, Gosnell B (1992) Taste responses and preferences for sweet high-fat foods: evidence for opioid involvement. Physiol Behav 51(2):371–379. https://doi.org/10.1016/0031-9384(92)90155-u

Flegal KM, Carroll MD, Ogden CL, Johnson CL (2002) Prevalence and trends in obesity among U.S. adults, 1999–2000. J Am Med Assoc 288(14):1723–1727. https://doi.org/10.1001/jama.288.14.1723

Food Marketing Institute, Better Homes and Gardens Magazine (1988) A study of food patterns and meal consumption. Food Marketing Institute, The Research Department, Washington, DC

Freedman D, Kettel Khan L, Serdula M, Srinivasan S, Berenson G (2001) BMI rebound, childhood height and obesity among adults: the Bogalusa heart study. Int J Obes 25(4):543–549. https://doi.org/10.1038/sj.ijo.0801581

Freedman DS, Khan LK, Serdula MK, Dietz WH, Srinivasan SR, Berenson GS (2004) Inter-relationships among childhood BMI, childhood height, and adult obesity: the Bogalusa heart study. Int J Obes 28(1):10–16. https://doi.org/10.1038/sj.ijo.0802544

Frijda N, Kuipers P, ter Schure E (1989) Relations among emotion, appraisal, and emotional action readiness. J Pers Soc Psychol 57(2):212–228. https://doi.org/10.1037/0022-3514.57.2.212

Galef BG Jr (1991) A contrarian view of the wisdom of the body as it relates to food selection. Psychol Rev 98(2):218–223. https://doi.org/10.1037/0033-295x.98.2.218

Gibney M, Sigman-Grant M, Stanton JL, Keast DR (1995) Consumption of sugars. Am J Clin Nutr 62(1):178S–194S. https://doi.org/10.1093/ajcn/62.1.178s

Gortmaker SL, Swinburn BA, Levy D, Carter R, Mabry PL, Finegood DT, Huang T, Marsh T, Moodie ML (2011) Changing the future of obesity: science, policy, and action. Lancet 378(9793):838–847. https://doi.org/10.1016/s0140-6736(11)60815-5

Goyal SK, Samsher GR, Goyal RK (2010) Stevia (Stevia rebaudiana) a bio-sweetener: a review. Int J Food Sci Nutr 61(1):1–10. https://doi.org/10.3109/09637480903193049

Harris G (2008) Development of taste and food preferences in children. Curr Opin Clin Nutr Metab Care 11(3):315–319. https://doi.org/10.1097/MCO.0b013e3282f9e228

Horton R (2018) Offline: NCDs, WHO, and the neoliberal utopia. Lancet 391(10138):2402. https://doi.org/10.1016/s0140-6736(18)31359-x

Houston DC (1979) The adaptation of scavengers. Serengeti: dynamics of an ecosystem. University of Chicago Press, Chicago, pp 263–286

Institute of Medicine, Food and Nutrition Board, Board on Children, Youth, and Families, Committee on Food Marketing and the Diets of Children and Youth, McGinnis M, Appleton Gootman J, Kraak VI (2006) Food marketing to children and youth: threat or opportunity? National Academies Press, Washington, DC

Jha P, Chaloupka FJ, Corrao M, Jacob B (2006) Reducing the burden of smoking world-wide: effectiveness of interventions and their coverage. Drug Alcohol Rev 25(6):597–609. https://doi.org/10.1080/09595230600944511

Kaplan HI, Kaplan HS (1957) The psychosomatic concept of obesity. J Nerv Ment Dis 125(2):191–201. https://doi.org/10.1097/00005053-195704000-00004

Kern DL, McPhee L, Fisher J, Johnson S, Birch LL (1993) The postingestive consequences of fat condition preferences for flavors associated with high dietary fat. Physiol Behav 54:71–76; 54(1):71–76. https://doi.org/10.1016/0031-9384(93)90045-h

Kloesel L (2000) Sugar substitutes. Int J Pharm Compd 4(2):86–87

Krause MV, Mahan LK (1984) Food, nutrition and diet therapy, 7th edn. WB Saunders Company, Philadelphia

Le Magnen J (1986) Hunger. Cambridge University Press, Cambridge

Lemus-Mondaca R, Vega-Gálvez A, Zura-Bravo L, Ah-Hen K (2012) Stevia rebaudiana Bertoni, source of a high-potency natural sweetener: a comprehensive review on the biochemical, nutritional and functional aspects. Food Chem 132(3):1121–1132. https://doi.org/10.1016/j.foodchem.2011.11.140

Levy DT, Mabry PL, Graham AL, Orleans CT, Abrams DB (2010) Reaching healthy people 2010 by 2013: a SimSmoke simulation. Am J Prev Med 38(3):S373–S381. https://doi.org/10.1016/j.amepre.2009.11.018

Logue AW (1986) The psychology of eating and drinking. Freeman, New York

Macht M (1999) Characteristics of eating in anger, fear, sadness, and joy. Appetite 33(1):129–139. https://doi.org/10.1006/appe.1999.0236

McCann BS, Warnick GR, Knopp RH (1990) Changes in plasma, lipids, and dietary intake accompanying shifts in perceived work-load and stress. Psychosom Med 52(1):97–108. https://doi.org/10.1097/00006842-199001000-00008

McCloy J, McCloy RF (1979) Enkephalin, hunger and obesity. Lancet 314(8134):156. https://doi. org/10.1016/s0140-6736(79)90047-3

Mennella JA, Beauchamp GK (1996) Taste and smell perception during human infancy. In: Calpaldi ED (ed) Why we eat what we eat: the psychology of eating. American Psychological Association, Washington, DC, pp 83–112

Mistretta CM (1981) Neurophysiological and anatomical aspects of taste development. In: Ash RN, Alberts JR, Peterson MR (eds) Development of perception, vol 1. Academic Press, New York, pp 433–455

Oliver G, Wardle J (1999) Perceived effects of stress on food choice. Physiol Behav 66(3):511–515. https://doi.org/10.1016/s0031-9384(98)00322-9

Parisi S (2002) Profili evolutivi dei contenuti batterici e chimico-fisici in prodotti lattiero-caseari. Ind Aliment 41(412):295–306

Parisi S (2003) Evoluzione chimico-fisica e microbiologica nella conservazione di prodotti lattiero - caseari. Ind Aliment 42(423):249–259

Parisi S (2012) Food packaging and food alterations. The user-oriented approach. Smithers Rapra Technology Ltd., Shawbury

Parisi S (2013) Food industry and packaging materials. User-oriented guidelines for users. Smithers Rapra Technology Ltd., Shawbury

Perissé J, Sizaret F, François P (1969) The effect of income and the structure of the diet. FAO nutrition newsletter vol 7. Food and Agriculture Organization of the United Nations (FAO), Rome, pp 1–9

Phillips KC (1987) Stevia: steps in developing a new sweetener. In: Grenby TH (ed) Developments in sweeteners. Elsevier New York, p 1.5

Pliner P, Melo N (1997) Good neophobia in humans: effects of manipulated arousal and individual differences in sensation seeking. Physiol Behav 61(2):331–335. https://doi.org/10.1016/s0031-9384(96)00406-4

Popkin BM (1994) The nutrition transition in low-income countries: an emerging crisis. Nutr Rev 52(9):285–298. https://doi.org/10.1111/j.1753-4887.1994.tb01460.x

Popkin BM, Paeratakul S, Zhai F, Ge K (1995) Dietary and environmental correlates of Obesity in a population study in China. Obes Res 3(S2):135s–143s. https://doi.org/10.1002/j.1550-8528. 1995.tb00456.x

Powell LM, Chaloupka FJ (2009) Food prices and obesity: evidence and policy implications for taxes and subsidies. Milbank Q 87(1):229–257. https://doi.org/10.1111/j.1468-0009.2009.00554. x

Putnam JJ, Allshouse JE (1970) Food consumption, prices and expenditures, 1970–97. Statisti-cal bulletin no. 965, April 1999. United States Department of Agriculture, Economic Research Service, Food and Rural Economics Division, Washington, DC

Rolls BJ (1986) Sensory-specific satiety. Nutr Rev 44(3):93–101. https://doi.org/10.1111/j.1753-4887.1986.tb07593.x

Rolls BJ (1999) Do chemosensory changes influence food intake in the elderly? Physiol Behav 66(2):193–197. https://doi.org/10.1016/s0031-9384(98)00264-9

Rolls BJ, McDermott M (1991) Effects of age on sensory-specific satiety. Am J Clin Nutr 54(6):988–996. https://doi.org/10.1093/ajcn/54.6.988

Rowe RC, Sheskey PJ, Owen SC (eds) (2006) Handbook of pharmaceutical excipient, 5th edn. The Royal Pharmaceutical Society of Great Britain, London, and the American Pharmacists Association, Washington, DC

Sacks G, Veerman L, Moodie M, Swinburn B (2010) 'Traffic-light' nutrition labelling and 'junk-food' tax: a modelled comparison of cost-effectiveness for obesity prevention. Int J Obes (London) 35(7):1001–1009. https://doi.org/10.1038/ijo.2010.228

Sanjur D (1982) Social and cultural perspectives in nutrition. Prentice-Hall, Englewood Cliffs, p 336

Schmidt R, Janssen E, Haussler O, Duriez X, Baron R (2000) Evaluating toothpaste sweetening. Cosmet Toilet 115:49–53

Sclafani A (1991) Starch and sugar tastes in rodents: an update. Brain Res Bull 27(3–4):383–386. https://doi.org/10.1016/0361-9230(91)90129-8

Slochower J (1983) Excessive eating: the role of emotions and environment. Human Sciences Library, New York

Stanton MV, Robinson JL, Kirkpatrick SM, Farzinkhou S, Avery EC, Rigdon J, Offringa LC, Trepanowski JF, Hauser ME, Hartle JC, Cherin RJ, King AC, Ioannidis JP, Desai M, Gardner CD, Gardner CD (2017) DIETFITS study (diet intervention examining the factors interacting with treatment success)—study design and methods. Contemp Clin Trials 53:151–161. https://doi.org/10.1016/j.cct.2016.12.021

Stewart J, De Wit H, Eikelboom R (1984) Role of unconditioned and conditioned drug effects in the self-administration of opiates and stimulants. Psychol Rev 91(2):251–268. https://doi.org/10.1037/0033-295x.91.2.251

Sundaresan K (2018) Sweetening strategies. Prepared Foods 187(9):60–64

Swinburn B, Sacks G, Hall KD, McPherson K, Finegood DT, Moodie M, Gortmaker S (2011) The global obesity pandemic: shaped by global drivers and local environments. Lancet 378(9793):804–814. https://doi.org/10.1016/s0140-6736(11)60813-1

Swithers SE, Davidson TL (2008) A role for sweet taste: calorie predictive relations in energy regulation by rats. Behav Neurosci 122(1):161–173. https://doi.org/10.1037/0735-7044.122.1.161

Szpylka J, Thiex N, Acevedo B, Albizu A, Angrish P, Austin S, Bach Knudsen KE, Barber CA, Berg D, Bhandari SD, Bienvenue A, Cahill K, Caldwell J, Campargue C, Cho F, Collison MW, Cornaggia C, Cruijsen H, Das M, De Vreeze M, Deutz I, Donelson J, Dubois A, Duchateau GS, Duchateau L, Ellingson D, Gandhi J, Gottsleben F, Hache J, Hagood G, Hamad M, Haselberger PA, Hektor T, Hoefling R, Holroyd S, Holt DL, Horst JG, Ivory R, Jaureguibeitia A, Jennens M, Kavolis DC, Kock L, Konings EJM, Krepich S, Krueger DA, Lacorn M, Lassitter CL, Lee S, Li H, Liu A, Liu K, Lusiak BD, Lynch E, Mastovska K, McCleary BV, Mercier GM, Metra PL, Monti L, Moscoso CJ, Narayanan H, Parisi S, Perinello G, Phillips MM, Pyatt S, Raessler M, Reimann LM, Rimmer CA, Rodriguez A, Romano J, Salleres S, Sliwinski M, Smyth G, Stanley K, Steegmans M, Suzuki H, Swartout K, Tahiri N, Ten Eyck R, Torres Rodriguez MG, Van Slate J, Van Soest PJ, Vennard T, Vidal R, Hedegaard RSV, Vrasidas I, Vrasidas Y, Walford S, Wehling P, Winkler P, Winter R, Wirthwine B, Wolfe D, Wood L, Woollard DC, Yadlapalli S, Yan X, Yang J, Yang Z, Zhao G (2018) Standard method performance requirements (SMPRs®) 2018.001: sugars in animal feed, pet food, and human food. J AOAC Int 101(4):1280–1282. https://doi.org/10.5740/jaoacint.smpr2018.001

Tuomisto T, Tuomisto MT, Hetherington MM (1998) Reasons for initiation and cessation of eating in obese men and women and the affective consequence of eating in everyday situations. Appetite 30(2):211–222. https://doi.org/10.1006/appe.1997.0142

USDA (1995) Nutrition & your health: dietary guidelines for Americans. Home and garden bulletin No. 232. United States Department of Agriculture (USDA), Washington, DC. Available https://health.gov/dietaryguidelines/dga95/. Accessed 30 Oct 2018

Vos T, Carter R, Barendregt J, Mihalopoulos C, Veerman JL, Magnus A, Cobiac L, Bertram MY, Wallace AL, ACE–Prevention Team (2010) Assessing Cost-Effectiveness in Prevention (ACE–Prevention): Final report. University of Queensland, Brisbane and Deakin University, Melbourne

Wansink B (1994) Antecedents and mediators of eating bouts. Fam Consum Sci Res J 23(2):166–182. https://doi.org/10.1177/1077727x94232005

Wansink B (2002) Changing eating habits on the home front: lost lessons from World War II research. J Public Policy Mark 21(1):90–99. https://doi.org/10.1509/jppm.21.1.90.17614

Wansink B (2003) Profiling nutritional gatekeepers: three methods for differentiating influential cooks. Food Qual Prefer 14(4):289–297. https://doi.org/10.1016/s0950-3293(02)00088-5

Wansink B, Sudman S (2002) Consumer panels, 2nd edn. American Marketing Association, Chicago

Wansink B, Sonka ST, Cheney MM (2002) A cultural hedonic framework for increasing the consumption of unfamiliar foods: soy acceptance in Russia and Columbia. Rev Agric Econ 24(2):353–365. https://doi.org/10.1111/1467-9353.00102

Wansink B, Cheney MM, Chan N (2003) Exploring comfort food preferences across age and gender. Physiol Behav 79(4–5):739–747. https://doi.org/10.1016/s0031-9384(03)00203-8

Weingarten HP, Kulikovsky OT (1989) Taste-to-postingestive consequence conditioning: is the rise in sham feeding with repeated experience a learning phenomenon? Physiol Behav 45(3):471–476. https://doi.org/10.1016/0031-9384(89)90060-7

Wise R (1988) The neurobiology of craving: implications for the understanding and treatment of addiction. J Abnorm Psychol 97(2):118–132. https://doi.org/10.1037/0021-843x.97.2.118

Chapter 3
Fat Content in Selected Industrial Products. The Role of Selected Vegetable Oils

Abstract The evolution of dietary lifestyles in modern society has been promoted by the concomitant creation of new and improved food and beverage products. It can be assumed that the birth and growth of anthropic activities related to food production (agriculture and husbandry) have altered dietary lifestyles, in Western countries at least. The alteration could be considered as the modification of certain nutritional features, including the assumption of nutritive sugars, fibres, and the qualitative and quantitative profile of general foods and beverages when speaking of nutritive compounds such as fatty acids, macro- and micronutrients. With exclusive relation to fats, the production of vegetable oils and hydrogenated products as replacers for animal fats can be considered as a subsector of the entire chain of food production. Advantages include improved food palatability, cheap prices, possible options for different vegetables, the creation and development of new products such as mayonnaise and imitation cheeses, and technological improvement. On the other side, the demonstrated correlation with eating disorders, malnutrition, obesity, and general cardiovascular diseases has to be considered (with other factors). These aspects are discussed in detail, including a practical example: an analogue cheese with different animal-to-vegetable oil ratios.

Keywords Butter · Cottonseed oil · Imitation cheese · Margarine · Milk · Palm oil · Soybean oil

Abbreviations

DIETFITS	Diet Intervention Examining the Factors Interacting with Treatment Success
FAO	Food and Agriculture Organization of the United Nations
PUFA	Polyunsaturated fatty acid
RSPO	Roundtable on Sustainable Palm Oil
USDA	United States Department of Agriculture

© The Author(s), under exclusive license to Springer Nature Switzerland AG 2019   31
S. D. Sharma and M. Barone, *Dietary Patterns, Food Chemistry and Human Health*,
Chemistry of Foods, https://doi.org/10.1007/978-3-030-14654-2_3

3.1   Fat Vegetables as Lipid Surrogates in Food Products. Why?

The evolution of dietary lifestyles in modern society has been promoted by the concomitant creation of new and improved food and beverage products. This affirmation does not concern recent periods of the human history: on the contrary, it can be assumed that the birth and growth of anthropic activities related to food production (agriculture and husbandry) have slowly and profoundly modified the human diet worldwide. On the other hand, it may be also considered that recent developments of the production of food and feed products have rapidly altered dietary lifestyles, in Western countries at least. Basically, the alteration could be considered as the modification of certain nutritional features, including (Cordain et al. 2005):

(1) The assumption of nutritive sugars with the consequent glycemic disorders (Liu and Willett 2002)
(2) The qualitative and quantitative profile of general foods and beverages when speaking of nutritive compounds such as fatty acids (Institute of Medicine of the National Academies 2002; Jenkins et al. 1981; Simopoulos 2002; Spady et al. 1993; Stamler et al. 2000; Thorburn et al. 1987)
(3) The qualitative and quantitative profile of general foods and beverages when speaking of macronutrients and micronutrients (Johnston et al. 2004; Wald et al. 2002)
(4) The amount of fibres in the diet (Krauss et al. 2000).

In general, the relation between nutritive components of dietary patterns (not initially available in the pre-agricultural history) appears in favour of cereals, when speaking of ratio between different caloric contents. In detail, whole and refined cereals can approximately supply 2.3 times more calories than the heterogeneous group of dairy products (butter, milk, cheese, etc.), based on the re-elaboration of recently reported data (Cordain et al. 2005; Gerrior and Bente 2002; Tippett and Cleveland 2001; USDA 2018). At the same time, the ratio between refined sugars and dairy foods is approximately 1.8:1, with a similar result when speaking of the comparison between refined vegetable oils and fats and dairy products (1.7:1). Alcohol has a negligible amount if compared with other representatives of the 'modern' food products. This simple comparison highlights the role of cereals in the initial anthropic activities, but also a certain importance for husbandry. However, the refining of basic products concerns sugars, vegetable oils and fats, and alcohol also. With the exception of the last beverage, mainly linked to the social and religious history of human communities, it has to be considered that sugar and fat/oil refining has received notable attention. After all, refined sugars and fats/oils have a more evident and interesting caloric amount, if compared with dairy foods.

In addition, the comparison between these food categories shows similar values, if compared with dairy products. In other terms, the importance of these food ingredients is approximately the same. It has been reported in Chap. 2 that the palatability of 'comfort foods' and snack products is mainly appreciated by large amounts of modern food consumers because of the concomitant presence of both refined sugars (sucrose, glucose, fructose, various syrups...) and fatty ingredients (Drewnowski 1987; Drewnowski and Greenwood 1983; Drewnowski and Holden-Wiltse 1992; Drewnowski et al. 1985, 1987; McCann et al. 1990; Oliver and Wardle 1999; Weidner et al. 1996). The addition of salt should be also considered in this ambit. Interestingly, the addition of fatty matters can enhance the sweetness of several dairy products, including cheeses, with the result that foods with a low or negligible sugar amount can appear 'sweet' because of high-fat contents (Barbieri et al. 2014; Parisi 2002, 2003; Parisi et al. 2006). However, because of the notable price of milk-derived fat matters in certain countries (economic reasons, possible fluctuations of milk availability in certain seasons) or logistic factors, the addition of non-animal fat may become important enough, including also dairy products (Mania et al. 2018).

In general, animal-derived fats are obtained from milk (different species) and meats. With relation to the last sector, the analytical profile of fatty acids generally shows the predominance of saturated fatty acids (fat depots in mammals represent the biological storage of excess food energy in subcutaneous and abdominal regions) if compared with polyunsaturated fatty acids and monounsaturated fatty acids (mainly found in non-adipose tissues). Interestingly, the main amount of saturated fatty acids and all trans-fatty acids has been repeatedly correlated with most important eating disorders and chronic diseases, and not recommended (Cordain et al. 2005; European Food Safety Authority 2004; Institute of Medicine of the National Academies 2002; Simopoulos 2002). This situation is also evident when speaking of the predominance of these fatty acids in the most important representatives of Western dietary patterns, including at least: fatty meat preparations; cheeses; milk; butter; margarine; and baked foods (as the representative for cereals). In these foods, saturated fatty acids are prevailing, while essential polyunsaturated fatty acids and ω-3/ω-6 fatty acids are reduced, although their consumption is recommended by the Food and Agriculture Organization of the United Nations (FAO 2010). Different diet models have been proposed so far, including 'the Drinking Man's Diet', the Atkins Diet, the Paleo Diet, and the more recent ketogenic diet (Anthony 2018).

Apparently, the introduction of vegetable oils and fats in these products (e.g. muffins) should be considered with favour because of the increasing amount of polyunsaturated fatty acids (PUFA) in comparison with saturated fatty acids, although a certain augment of ω-6 PUFA has been reported against more preferred ω-3 PUFA, surpassing 9:1 ratios in the USA differently from reported results concerning 'old' foods (Cordain et al. 2000, 2002, 2005; Kris-Etherton et al. 2000). In addition, the use of vegetable oils and fats for palatability purposes has constantly increased the amount of total fat contents in prepared foods, especially after the production of the first margarine products by hydrogenation of vegetable oils. The process of hydrogenation allows the production of fat surrogates with good yields; on the other hand, the increase of an undesired fatty acid, *trans*-elaidic acid, has been reported and correlated with cardiovascular diseases in the USA and in other Western countries (Allison et al. 1999; Ascherio et al. 1994; Emken 1984). The amount of vegetable fats (if compared with normally contained animal fats) in certain imitation pizza or analogue cheeses (Mania et al. 2018; Parisi 2002, 2003) is one of the best-known evidence in the sector of industrial cheeses.

In general, vegetable oils and fats—with the exclusion of fine oils such as olive oil (Delgado et al. 2017)—can be used in different contexts: the most important of realised food products include baked foods, processed cheeses, and stand-alone products (mayonnaise, other food ingredients such as 'shortenings'). The continuous augment of the use of vegetable fats and oils in the industry has been globally confirmed in the Mediterranean basin, both in Western-style countries and in other countries where the 'Mediterranean Diet' is the most representative dietary pattern (da Silva et al. 2009; Keys et al. 1980; Vareiro et al. 2009). In detail, the consumption of cereals, alcoholic beverages, and legumes has progressively decreased, while milk, meat products, sugars, and vegetable oils (except for olive oils) have been reported to show increased consumption (Delgado et al. 2017). A recent research—the 'Diet Intervention Examining the Factors Interacting with Treatment Success' (DIETFITS)—has confirmed that genetic factors are not responsible for different answers—and the result—when speaking of low-fat or low-carbohydrate diets (Stanton et al. 2017).

Economic reasons should be considered when speaking of the main reason which vegetable fats and oils are used for. The right question is 'Why vegetable lipids as food additives'? However, technological factors should be also discussed, especially in relation to certain products and their peculiarities.

3.2 Fat Vegetables as Lipid Surrogates in Food Products. Why not?

First of all, the production of vegetable oils and hydrogenated products can be considered as a subsector of the entire chain of food production. In other words, certain raw materials could be used 'as they are' (e.g. olives) or be used as the starting point of another fine product(s). The best examples are—in a general way—milk (for the production of yogurts, cheeses, aromatised, or normal butter, etc.) and olives (for the production of different olive oils). Anyway, refined vegetable oils imply the use of selected vegetables, including some surprise (Delgado et al. 2017):

(a) Palm oil from palm trees
(b) Coco oils from coco nuts
(c) Mixtures of vegetable palm and coco oils
(d) Sunflower oil
(e) Colza oil from rapeseeds
(f) Safflower oils from Safflower (*Carthamus tinctorius*)
(g) Peanut oil from peanuts (pulses)
(h) Soybean oil from soybean seeds (pulses)
(i) Cottonseed oil from seeds of cotton plants.

Consequently, refined vegetable oils can represent an important 'slice' of the market and economic activities for Nations, including developing countries. On the other side, environmental concerns and ethical problems can be observed. In this ambit, the promotion of 'sustainable palm oil' by means of the Roundtable on Sustainable Palm Oil (RSPO)[1] would defend the production of palm oil reducing environmental and social negative effect of the cultivation of palm trees.

Secondly, vegetable oils are the basis for the production of a well-known stand-alone product, mayonnaise (Duncan 2008). This food is a semisolid emulsion obtained from vegetable oils, vinegar, egg yolk, and some vegetable juice (lime or lemon) and egg yolk. Interestingly, the composition of this product should guarantee at least 65% of vegetable oil(s) in weight; the list of allowed additives contains some limitations, especially when speaking of spices visually able to simulate yellow colours typical of egg yolk (Duncan 2008). The example of mayonnaise is good enough because many of the above-mentioned vegetable oils can be used in the formulation: soybean, safflower, corn, olive, and cottonseed oils are reported in this ambit (Depree and Savage 2001; Duncan 2008; Krishnamurthy and Witte 1996). On the other hand, it should be mentioned that saturated oils such as palm oil or each type of vegetable oil able to solidity in chillers could be the cause of emulsion break at low temperatures.

Another example of the variegated use of vegetable oils is represented by shortenings for pastries and baked foods. In summary, this ingredient is used with the aim of making plastic masses or enhancing the natural plasticity of intermediate masses for baked products. The obtained result is the augment of volumes by means of air incorporation. Also, these shortenings can be used as frying fats in many applications. Because of their large applications, the industry of baked foods could not exclude the use of similar products obtained from the mixture of soybean oil with cottonseed oil (although soybean oil can be the only constituent). The presence of impurities has to be avoided strictly, for this reason, free fatty acids have to be saponified with alkali; also, bleaching treatments are required because of the possible presence of residual xanthophylls, carotenes, or other vegetable pigments able to give greyish or greenish colours (Carden and Basilio 2008). As explained above, hydrogenation is preferred when speaking of products with a semisolid texture and with the ability to solidify at cold temperatures: raw materials used for shortenings are not the exception to this 'rule'.

The example of shortenings can explain various advantages of the use of similar products by the technological angle. These cheap additives and vegetable oils themselves can be used in the production of baked products—muffins, bread—for three essential and technological reasons:

(a) Vegetable oils can augment (Bechtel et al. 1978) the volumetric expansion of the flour paste (bread: 15–25%).

(b) The presence of vegetable additives is greatly appreciated in certain baked products (muffins) where the desired amount of fat matter exceed 18% (in muffins, this content is between 18 and 40%). The reason is the enhancement of crustiness and soft texture. Also, the fat components help to retain water, with additional durability performances (McWilliams 2001).

---

[1] Detailed information can be found online at the following web site: www.rspo.org.

(c) Aroma is greatly appreciated when using vegetable fats. It has been reported that baked products have a great performance in this ambit because odorous molecules are easily dissolved in the fat phase.

Because of some reliable health worrying related to the consumption of these products because of the high amount of saturated fatty acids, alternative solutions have recently concerned the use of PUFA-rich shortenings with 13% of safflower oil (Berglund and Hertsgaard 1986; Cross 2008).

In addition, the presence of fat matters—including vegetable fats and oils—in certain products has been progressively reported to have good effects when speaking of microwave heating (Fu 2008). In detail, deep-frozen foods are known to show several textural problems when defrosted, because frozen water is not able to exhibit microwave dipole relaxation while frozen products contain also liquid water. As a result, microholes or micro- /macroscopic fractures can be observed. The presence of fat globules, able to exhibit low dielectric losses, allows heating the whole product in a rapid way, with the consequent minimisation of textural defects (Fu 2008).

As a result, the use of selected vegetable oils as food additives and surrogates may be evaluated in the following way:

(a) Advantages: improved food palatability; cheap prices if compared with 'animal' competitor raw materials; possible options for different vegetables (in terms of sub- and by-products such as food surrogates); creation and development of new products such as mayonnaise; development of new products containing refined vegetable oils, such as imitation cheese; technological properties (improvement of texture, crustiness, cooking performances, volumetric expansion, flavour, sweet taste if in conjunction with sugars); possible realisation of vegetable oils containing more PUFA and monounsaturated fatty acids than saturated and all *trans*-fatty acids; reduction of cholesterol values in cheese containing animal fat

(b) Risks: demonstrated correlation with eating disorders, malnutrition, obesity, and general cardiovascular diseases; damages to the environment and social communities in certain situations (for developing countries); realisation of products with higher fat content if compared with original (traditional or historical) recipes or versions, with the consequent increase of dietary intakes.

The above-mentioned list of advantages and risks may be not completely exhaustive. Anyway, the presence of refined vegetable fats and oils is a constant of the whole world of foods and beverages.

A short analysis and description of each vegetable oil or fat typology could be very challenging. This chapter would give a brief description of a type of vegetable fat for a detailed production only, with the aim of exploring current possibilities: refined palm oil for the production of imitation cheeses (Mania et al. 2018; Parisi 2002, 2003).

3.3 Fat Vegetables as Lipid Surrogates in Food Products. Palm Oil for Processed Cheeses

Palm oil is one of the most used vegetable materials for fat substitution (instant noodles, processed cheeses, etc.) and food-related operations such as industrial frying, after chemical treatments such as esterification (Arslan et al. 2010; Dobraszczyk et al. 2006; Norizzah et al. 2004). At present, Malaysia and Indonesia and the most represented countries when speaking of crude palm oil production (Belitz et al. 2009).

The main feature of this material is the ready solidification at low temperatures. Substantially, the original fat matter is extracted as a semisolid material and subsequently separated into two main fractions:

(1) Liquid olein (melting point between 19 and 24 °C—useful for frying applications and fat replacement in cheeses, although some additional refining for better rheological properties is needed), and
(2) The more solid stearin (melting point $\geq$44 °C). It should be removed.

With relation to normal uses, it has to be noted that refined palm oils for cheese productions should not exhibit dissimilar properties if compared with animal butter. As a result, rheological values, peroxide values, and so on, do not differ from the usual butter. However, the main problem of all palm oils is always correlated with the natural abundance of palmitic (saturated) fatty acid (a C16:0 chain, melting point: 63 °C), typically higher than 40.0% and lower than 47.0%), and with health worries (Dobraszczyk et al. 2006; Pantzaris 1999). The remaining fat fraction shows the high content of oleic acid (a C18:1 unsaturated chain, melting point: 16 °C), if compared with other C12–C18 fatty acids. Oleic acid is approximately 39%; the C14 fraction 'weights' 1% only, where linoleic acid, C18:2, is approximately 10% (Belitz et al. 2009). Also, saturated fatty acids naturally influence rheological properties, including technological applications for cheese processing.

The presence of carotenoids, determined analytically as a percentage of β-carotene by means of spectrophotometric absorption of fatty solutions in cyclohexane (λ = 445 nm), may be important enough because of the visual appearance of the final product (FSSAI 2015). In certain situations, yellow colours may be useful (as colourant substances for margarine), but refined oils show normally white-to-yellow colours. The normal original amount of α- and β-carotenes is not higher than 0.2% (Belitz et al. 2009).

In fact, crude oils have to be refined with the aim of removing all possible non-fat impurities. The sequence of operations concerns (in general, for all vegetable oils including palm oil) the following steps (Belitz et al. 2009; Nawar 1996):

(a) Degumming (at 50% with water, if the crude oils also contain phospholipids).
(b) Settling (at 50 °C: the oil matter separates itself from the aqueous phase). A centrifugation is needed after the removal of water, protein-like matters, carbohydrates, etc., phospholipids, and carbohydrates.
(c) Washing with hot water (carotenes are concentrated).

(d) Neutralisation (removal of free fatty acid, residual phospholipids, and various pigments). Sodium hydroxide is needed.
(e) Bleaching (subsequent removal of impurities, including carotenes, with heat treatments).
(f) Deodorisation (by steam distillation).
(g) Hydrogenation (the production of semisolid fat matters is possible in this way). Certain palm oils for cheese production do not contain hydrogenated oils and fats, with consequent reduction of solidifying performances. Fully hydrogenated palm oil shows palmitic and stearic acids (a C18:0 fatty acid, melting point: 70 °C) correspond to 97% of the whole fat mass.
(h) Interesterification at high temperatures with palm seed, or coconut oils, and catalysers (the initial and non-random distribution of fatty acids in fat molecules is randomly modified, with the improvement of oil properties.

One of the main applications of refined palm oils as the solid material is the use as a fat replacer in the so-called imitation cheeses: products obtained from thermally melted mixtures of rennet casein, vegetable fats, animal fats (butter), water, and some additive such as citric acid or trisodium citrate. The important point is that this fat matter can partially or totally replace the original and supposed fat content of the cheese (original cheeses should be intended as 'fat preserves'), with interesting cheap prices (Javidipour and Tuncturk 2007; Mania et al. 2018; Parisi 2002, 2003; Parisi and Luo 2018). The reduction of these prices is correlated with the increasing amount of vegetable fats in the formula, usually between 22 and 28% (Guinee 2007a).

A peculiar defect in some of these imitation cheeses is soapiness, determined probably by sodium or potassium oleate obtained from emulsion agents containing sodium and potassium on the one hand, and oleic acid (the higher the fat content, as in imitation cheeses, the higher the defect). The problem is not frequent, but the reason is that different raw materials—animal and vegetable compounds—have to be mixed and remain with a stable structure and texture. In fact, too low-fat content may because of the production of too dry or crumbly cheeses (Guinee 2007b, c; Guinee et al. 2004). Another interesting property is the ability of obtained cheeses to dry in a few minutes only at 350 °C and higher temperatures in oven; pizza obtained with imitation cheeses show a melted cheese layer with improved rigidity and reduced superficial water (while a part of oils is normally expelled) with enhanced performances in terms of meltability and time savings.

The use of palm oils may be also useful when speaking of microwave heating and processed cheeses. These products in the deep-frozen version may show several textural problems when defrosted (frozen water is not able to exhibit microwave dipole relaxation while frozen products contain also liquid water). Figures 3.1 and 3.2 show two different processed kinds of cheese obtained by means of:

(1) The addition of butter (15%) and palm oil (5%), and
(2) The addition of butter (15%) and palm oil (15%).

Figure 3.1 shows some defects after a light microwave defrosting (several seconds only) before heavy melting (normal and dark photographs), while Fig. 3.2 shows

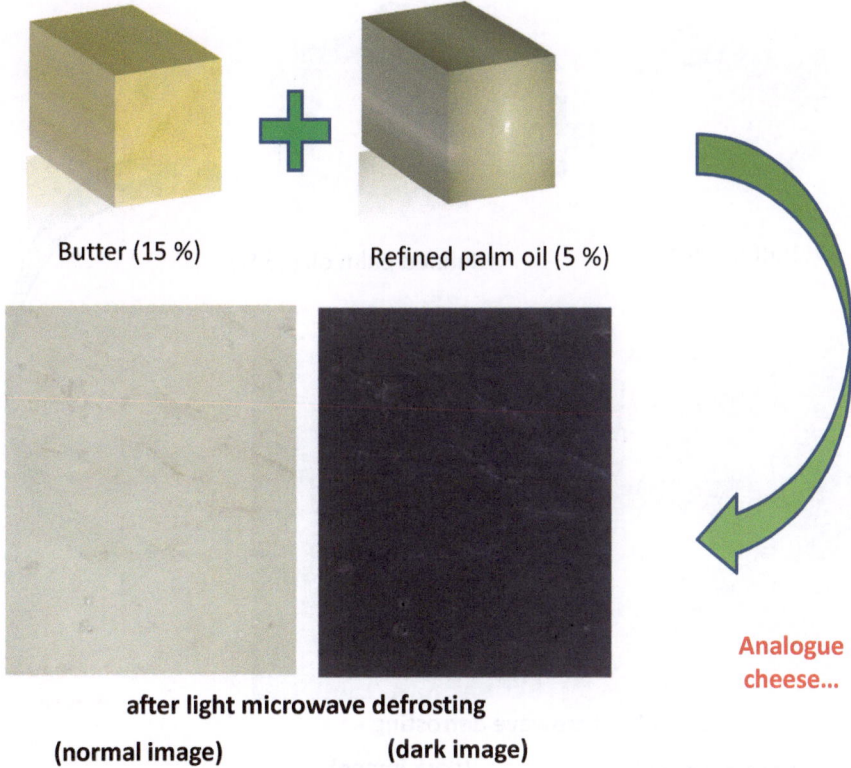

Butter (15 %)   **Refined palm oil (5 %)**

after light microwave defrosting

(normal image)   **(dark image)**

Analogue cheese…

Fig. 3.1 Use of palm oils in processed cheeses may have some advantages when speaking of frozen products. An analogous cheese obtained with the addition of butter (15%) and palm oil (5%) can show some superficial defects (normal and dark photographs) after a light microwave defrosting (several seconds only) before heavy melting. Figure 3.2 shows the effect of different formulations with relation to the fat matter

the second product with reduced fractures (after the same microwave defrosting). Actually, the minimisation of defects should be examined in terms of rapid heating; the cause is essentially the augment of the total fat content (from $15\% + 5\% = 20\%$ to $15\% + 15\% = 30\%$) without relation to the type of animal or vegetable fat. However, being palm oil cheaper than butter, the second result (Fig. 3.2) is economically and technologically preferred.

A problem concerning palm oil is the amount of cholesterol. Before refining, the related amount is only 2.8%; after refining, present glycosides are altered, with the augment of detectable cholesterol up to 8.8% (Belitz et al. 2009).

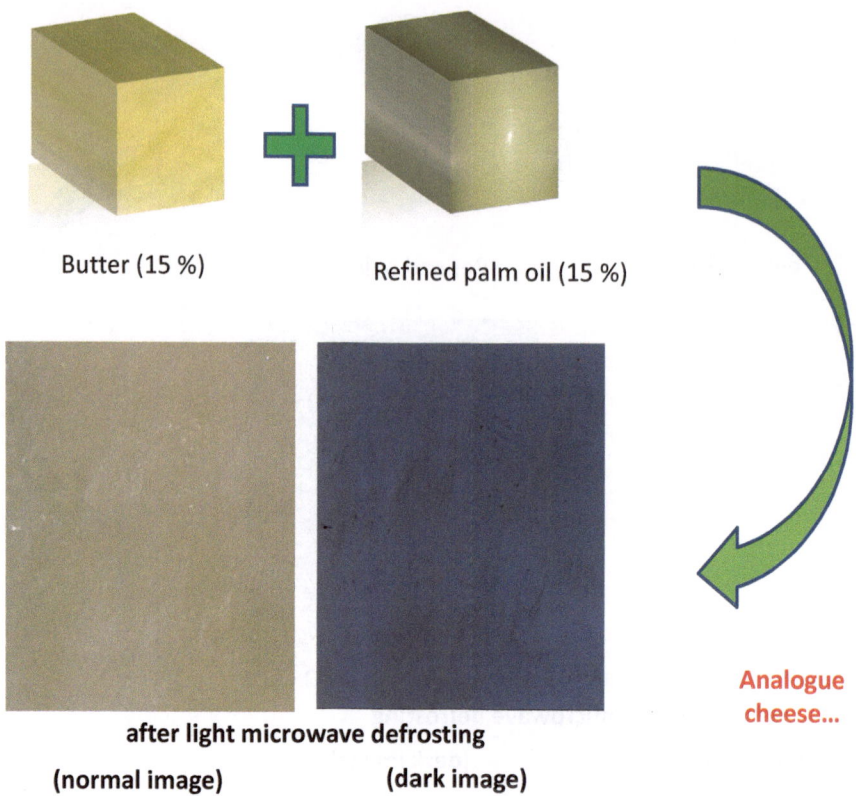

Butter (15 %)   **Refined palm oil (15 %)**

after light microwave defrosting

(normal image)   **(dark image)**

Analogue cheese...

Fig. 3.2 Use of palm oils in processed cheeses may have some advantages when speaking of frozen products. An analogous cheese obtained with the addition of butter (15%) and palm oil (15%) can show light superficial defects (normal and dark photographs) after a light microwave defrosting (several seconds only) before heavy melting. The comparison with another melted cheese obtained with the addition of butter (15%) and palm oil (5%) as displayed in Fig. 3.1 might show that the higher the amount of vegetable fat, the lower the presence of superficial fractures. However, the minimisation of defects should be examined in terms of rapid heating; the cause is essentially the augment of the total fat content (from $15\% + 5\% = 20\%$ to $15\% + 15\% = 30\%$) without relation to the type of animal or vegetable fat. However, being palm oil cheaper than butter, the augment of vegetable oils is economically and technologically preferred

References

Allison DB, Egan SK, Barraj LM Caughman C, Infante M, Heimbach JT (1999) Estimated intakes of trans fatty and other fatty acids in the US population. J Am Diet Assoc 99(2):166–174. https://doi.org/10.1016/s0002-8223(99)00041-3

Almeida MDV, Parisi S, Delgado AM (2017) Food and nutrient features of the Mediterranean DieT. In: Delgado A, Vaz de Almeida MD, Parisi S (2016) Chemistry of the Mediterranean diet. Springer International Publishing, Cham. https://doi.org/10.1007/978-3-319-29370-7_2

Anthony M (2018) Diet is a four-letter word. Prepared Foods 18(9):45–51

Arslan S, Topcu A, Saldamli I, Koksal G (2010) Utilization of interesterified fat in the production of Turkish white cheese. Food Sci Biotechnol 19(1):89–98. https://doi.org/10.1007/s10068-010-0013-2

Ascherio A, Hennekens CH, Buring JE Master C, Stampfer MJ, Willett WC (1994) Trans-fatty acids intake and risk of myocardial infarction. Circulation 89(1):94–101. https://doi.org/10.1161/01.cir.89.1.94

Barbieri G, Barone C, Bhagat A, Caruso G, Conley ZR, Parisi S (2014) The influence of chemistry on new foods and traditional products. Springer International Publishing, Cham

Bechtel DB, Pomeranz Y, deFrancisco A (1978) Breadmaking studied by light and transmission electron microscopy. Cereal Chem 55(3):392–401

Belitz HD, Grosch W, Schieberle P (2009) Food chemistry, fourth revised and extended edition. Springer, Berlin

Berglund PT, Hertsgaard DM (1986) Use of vegetable oils at reduced levels in cake, pie crust, cookies, and muffins. J Food Sci 51(3):640–644. https://doi.org/10.1111/j.1365-2621.1986.tb13899.x

Carden LA, Basilio LK (2008) Fats: vegetable shortening. In: Scott Smith J, Hui YH (eds) Food processing: principles and applications. Blackwell Publishing, Ames

Cordain L, Brand Miller J, Eaton SB, Mann N, Holt SHA, Speth JD (2000) Plant to animal subsistence ratios and macronutrient energy estimations in world wide hunter-gatherer diets. Am J Clin Nutr 71(3):682–692. https://doi.org/10.1093/ajcn/71.3.682

Cordain L, Eaton S, Miller JB, Mann N, Hill K (2002) The paradoxical nature of hunter-gatherer diets: meat-based, yet non-atherogenic. Eur J Clin Nutr 56(S1):S42–S52. https://doi.org/10.1038/sj.ejcn.1601353

Cordain L, Eaton SB, Sebastian A, Mann N, Lindeberg S, Watkins BA, O'Keefe JH, Brand-Miller J (2005) Origins and evolution of the Western diet: health implications for the 21st century. Am J Clin Nutr 81(2):341–354. https://doi.org/10.1093/ajcn.81.2.341

Cross N (2008) Bakery: muffins. In: Scott Smith J, Hui YH (eds) Food processing: principles and applications. Blackwell Publishing, Ames

da Silva R, Bach-Faig A, Quintana BR, Buckland G, Almeida MDV, Serra-Majem L (2009) Worldwide variation of adherence to the Mediterranean diet, in 1961–1965 and 2000–2003. Public Health Nutr 12(9A):1676–1684. https://doi.org/10.1017/S1368980009990541

Delgado A, Vaz de Almeida MD, Parisi S (2017) Chemistry of the Mediterranean diet. Springer International Publishing, Cham

Depree JA, Savage GP (2001) Physical and flavour stability of mayonnaise. Trends Food Sci Technol 12(5–6):157–163. https://doi.org/10.1016/s0924-2244(01)00079-6

Dobraszczyk BJ, Ainsworth P, Ibanoglu S, Bouchon P (2006) Baking, extrusion and frying. In: Brennan JG (ed) Food processing handbook. Wiley-VCH Verlag GmbH & Co. KGaA, Weinheim. https://doi.org/10.1002/3527607579.ch8

Drewnowski A (1987) Fats and food texture: sensory and hedonic evaluations. In: Moskowitz HR (ed) Food texture. Marcel Dekker, New York, pp 217–250

Drewnowski A (1997) Taste preferences and food intake. Annu Rev Nutr 17(1):237–253. https://doi.org/10.1146/annurev.nutr.17.1.237

Drewnowski A, Greenwood MR (1983) Cream and sugar: human preferences for high-fat foods. Physiol Behav 30(4):629–633. https://doi.org/10.1016/0031-9384(83)90232-9

Drewnowski A, Holden-Wiltse J (1992) Taste responses and food preferences in obese women: effects of weight cycling. Int J Obes 16:639–648

Drewnowski A, Brunzell JD, Sande K, Iverius PH, Greenwood MRC (1985) Sweet tooth reconsidered: taste responsiveness in human obesity. Physiol Behav 35(4):617–622. https://doi.org/10.1016/0031-9384(85)90150-7

Duncan SE (2008) Fats: Mayonnaise. In: Scott Smith J, Hui YH (eds) Food processing: principles and applications. Blackwell Publishing, Ames

Emken EA (1984) Nutrition and biochemistry of trans and positional fatty acid isomers in hydrogenated oils. Annu Rev Nutr 4(1):339–376. https://doi.org/10.1146/annurev.nu.04.070184.002011

European Food Safety Authority (2004) Opinion of the scientific panel on dietetic products, nutrition and allergies on a request from the commission related to the presence of trans fatty acids in foods and the effect on human health of the consumption of trans fatty acids. EFSA J 81:1–49. https://doi.org/10.2903/j.efsa.2004.81

FAO (2010) Fats and fatty acids in human nutrition: report of an expert consultation. FAO food and nutrition paper 9. Food and Agricultural Organization of the United Nations (FAO), Rome

FSSAI (2015) Manual of methods of analysis of foods—oils and fats. Food Safety and Standards Authority of India (FSSAI) (Ministry of Health and Family Welfare), New Delhi

Fu YC (2008) Applications of microwave and radio frequency in food processing. In: Scott Smith J, Hui YH (eds) Food processing: principles and applications. Blackwell Publishing, Ames

Gerrior S, Bente L (2002) Nutrient content of the U.S. food supply, 1909–99: a summary report. Home economics report No 55. United States Department of Agriculture (USDA), Center for Nutrition Policy and Promotion, Washington, DC

Guinee TP (2007a) Introduction: what are analogue cheeses? In: McSweeney PLH (ed) Cheese problems solved. CRC Press, Boca Raton, Boston

Guinee TP (2007b) Why does processed cheese have a dry, short, crumbly texture? In: McSweeney PLH (ed) Cheese Problems Solved. CRC Press, Boca Raton, Boston

Guinee TP (2007c) Why does processed cheese sometimes have a soapy flavour? In: McSweeney PLH (ed) Cheese problems solved. CRC Press, Boca Raton, Boston

Guinee TP, Carić M, Kaláb M (2004) Pasteurized processed cheese and substitute/imitation cheese products. In: Fox PF, McSweeney PLH, Cogan TM, Guinee TP (eds) Cheese: chemistry, physics and microbiology, vol 2, pp 349–394. https://doi.org/10.1016/s1874-558x(04)80052-6

Institute of Medicine of the National Academies (2002) Dietary fats: total fat and fatty acids. In: Dietary reference intakes for energy, carbohydrate, fiber, fat, fatty acids, cholesterol, protein, and amino acids (macronutrients). The National Academy Press, Washington, DC, pp 335–432. Available   https://www.nal.usda.gov/sites/default/files/fnic_uploads/energy_full_report.pdf. Accessed 31th Oct 2018

Javidipour I, Tuncturk YI (2007) Effect of using interesterified and noninteresterified corn and palm oil blends on quality and fatty acid composition of Turkish white cheese. Int J Food Sci Tech 42(12):1465–1474. https://doi.org/10.1111/j.1365-2621.2006.01366.x

Jenkins DJ, Wolever TM, Taylor RH, Barker H, Fielden H, Baldwin JM, Bowling AC, Newman HC, Jenkins AL, Goff DV (1981) Glycemic index of foods: a physiological basis for carbohydrate exchange. Am J Clin Nutr 34(3):362–366. https://doi.org/10.1093/ajcn/34.3.362

Johnston CS, Tjonn SL, Swan PD (2004) High-protein, low-fat diets are effective for weight loss and favorably alter biomarkers in healthy adults. J Nutr 134(3):586–591. https://doi.org/10.1093/jn/134.3.586

Keys A, Aravanis C, Blackburn I, Buzina R, Djordjevic BS, Dontas AS, Fidanza F, Karvonen MJ, Kimura N, Menotti A, Muhacek I, Nedeljkovic S, Puddu V, Punsar S, Taylor HL, van Buchem FSP (1980) Seven countries—a multivariate analysis of death and coronary heart disease. A commonwealth fund book. Harvard University Press, Cambridge

Krauss RM, Eckel RH, Howard B, Appel LJ, Daniels SR, Deckelbaum RJ, Erdman JW, Kris-Etherton P, Goldberg IJ, KotchenTA, Lichtenstein AH, Mitch WE, Mullis R, Robinson K, Wylie-Rosett J, St. Jeor S, Suttie J, Tribble DL, Bazzarre TL (2000) AHA dietary guidelines: revision 2000: a statement for healthcare professionals from the Nutrition Committee of the American Heart Association. Circulation 102(18):2284–2299. https://doi.org/10.1161/01.cir.102.18.2284

Kris-Etherton PM, Taylor DS, Yu-Poth S, Huth P, Moriarty K, Fishell V, Hargrove RL, Zhao G, Etherton TD (2000) Polyunsaturated fatty acids in the food chain in the United States. Am J Clin Nutr 71(1):179S–188S. https://doi.org/10.1093/ajcn/71.1.179s

Krishnamurthy RG, Witte VC (1996) Cooking oils, salad oils, and oil-based dressings. In: Hui YH (ed) Bailey's industrial oil and fat products, 5th edn. vol 3. Wiley, Hoboken, pp 193–223

Liu S, Willett WC (2002) Dietary glycemic load and atherothrombotic risk. Curr Atheroscler Rep 4(6):454–461. https://doi.org/10.1007/s11883-002-0050-2

Mania I, Delgado AM, Barone C, Parisi S (2018) Traceability in the dairy industry in Europe. Springer International Publishing, Cham

McCann BS, Warnick GR, Knopp RH (1990) Changes in plasma, lipids, and dietary intake accompanying shifts in perceived work-load and stress. Psychosom Med 52(1):97–108. https://doi.org/10.1097/00006842-199001000-00008

McWilliams M (2001) Fats and oils in food products. Foods: experimental perspectives, 4th edn. Prentice Hall, Upper Saddle River, pp 245–265

Nawar WW (1996) Lipids. In: Fennema OR (ed) Food chemistry, 3rd edn. Marcel Dekker, New York

Norizzah AR, Chong CL, Cheow CS, Zaliha O (2004) Effects of chemical interesterification on physicochemical properties of palm stearin and palm kernel olein blends. Food Chem 86(2):229–235. https://doi.org/10.1016/j.foodchem.2003.09.030

Oliver G, Wardle J (1999) Perceived effects of stress on food choice. Physiol Behav 66(3):511–515. https://doi.org/10.1016/s0031-9384(98)00322-9

Pantzaris TP (1999) Palm oil in frying. In: Boskoy D, Elmadfa I (eds) Frying of food. Technomic Publishing, Lancaster, pp 223–225

Parisi S (2002) Profili evolutivi dei contenuti batterici e chimico-fisici in prodotti lattiero-caseari. Ind Aliment 41(412):295–306

Parisi S (2003) Evoluzione chimico-fisica e microbiologica nella conservazione di prodotti lattiero - caseari. Ind Aliment 42(423):249–259

Parisi S, Luo W (2018) Chemistry of Maillard reactions in processed foods. Springer International Publishing, Cham

Parisi S, Laganà P, Delia S (2006) Curve di crescita dei miceti in diversi formaggi in atipiche condizioni di conservazione. Ind Aliment 45(458):532–538

Simopoulos AP (2002) Omega-3 fatty acids in inflammation and autoimmune disease. J Am Coll Nutr 21(6):495–505. https://doi.org/10.1080/07315724.2002.10719248

Spady DK, Woollett LA, Dietschy JM (1993) Regulation of plasma LDL-cholesterol levels by dietary cholesterol and fatty acids. Annu Rev Nutr 13(1):355–381. https://doi.org/10.1146/annurev.nu.13.070193.002035

Stamler J, Daviglus ML, Garside DB Dyer AR, Greenland P, Neaton JD (2000) Relationship of baseline serum cholesterol levels in 3 large cohorts of younger men to long-term coronary, cardiovascular, and all-cause mortality and to longevity. JAMA 284(3):311–318. https://doi.org/10.1001/jama.284.3.311

Stanton MV, Robinson JL, Kirkpatrick SM, Farzinkhou S, Avery EC, Rigdon J, Offringa LC, Trepanowski JF, Hauser ME, Hartle JC, Cherin RJ, King AC, Ioannidis JP, Desai M, Gardner CD, Gardner CD (2017) DIETFITS study (diet intervention examining the factors interacting with treatment success)—study design and methods. Contemp Clin Trials 53:151–161. https://doi.org/10.1016/j.cct.2016.12.021

Thorburn AW, Brand JC, Truswell AS (1987) Slowly digested and absorbed carbohydrate in traditional bushfoods: a protective factor against diabetes? Am J Clin Nutr 45(1):98–106. https://doi.org/10.1093/ajcn/45.1.98

Tippett KS, Cleveland LE (2001) Results from USDA's 1994–96 diet and health knowledge survey. NFS report No. 96-4, May 2001. United States Department of Agriculture (USDA), Agricultural Research Service, Washington, DC. Available https://www.ars.usda.gov/ARSUserFiles/80400530/pdf/dhks9496.PDF. Accessed 31th Oct 2018

USDA (2018) Food availability (Per Capita) data system. United States S Department of Agriculture (USDA), Economic Research Service. Food Consumption (per capita) data system. Available https://www.ers.usda.gov/data-products/food-availability-per-capita-data-system/. Accessed 31th Oct 2018

Vareiro D, Faig AB, Quintana B, Bertomeu I, Buckland G, Almeida MDV, Majem LS (2009) Availability of Mediterranean and non-Mediterranean foods during the last four decades: comparison of several geographical areas. Public Health Nutr 12(9A):936–941. https://doi.org/10.1017/S136898000999053X

Wald DS, Law M, Morris JK (2002) Homocysteine and cardiovascular disease: evidence on causality from a meta-analysis. BMJ 325(7374):1202–1206. https://doi.org/10.1136/bmj.325.7374.1202

Weidner G, Kohlmann CW, Dotzauer E, Burns LR (1996) The effects of academic stress on health behaviors in young adults. Anxiety Stress Coping 9(2):123–133. https://doi.org/10.1080/10615809608249396

Chapter 4
Protein Sources in the Modern Food Industry. Are Vegan Foods the Right Choice?

Abstract The current industry of foods and beverages has modified the market in recent years. One of the most interesting modifications concerning dietary patterns in Western Countries at least has been the emersion of so-called vegan foods. By the chemical viewpoint, the formulation of similar foods can be promising because many vegetable ingredients are good protein sources. The basic advantages of these productions—and vegan foods themselves—are: the absence of taste; eco-friendly methods for agriculture and industrial transformation; lower prices if compared with animal protein sources; possibility of new and improved 'surrogate' products; and improved technological features. In addition, veganism concerns one or more of the following motivations: ethical reasons; repugnance for food of animal origin; health reasons; explicit preference for vegetable/vegetarian foods and patterns; and social pressure in terms of persuasion. However, could these animal imitation foods give the 'right' amount of proteins to vegan and normal consumers? The problem of proteins should be discussed in terms of essential amino acids for the human being. Vegan-style diets may have good effects on certain patients; however, the energy intake related to proteins is significantly lower when non-vegetarian people. Also, the supplementation of some deficient vitamins, dietary fibres, and certain metals is recommended.

Keywords All *trans*-fatty acids · Iron · Soy protein · Veganism · Vegetarianism · Vitamin B_{12} · Zinc

4.1 The Evolution of Dietary Lifestyles. Vegan Foods

The current industry of foods and beverages has modified the market in recent years. One of the most interesting modifications concerning dietary patterns in Western Countries at least has been the emersion of so-called vegan foods. However, could these animal imitation foods give the 'right' amount of proteins to vegan (and normal) consumers? By the chemical viewpoint, the formulation of similar foods can be promising because many vegetable ingredients are good protein sources. This chapter would give a comprehensive overview of current options.

Basically, the increasing demand for vegetable—totally vegetable—foods has been justified by a low but important part of consumers worldwide (González et al. 2015) with the need of avoiding animal products for edible consumption (ethical concerns). In addition, the reduced consumption of animal foods is generally recommended for health reasons (Elmadfa and Freisling 2005; Kaplan 1984; Law 2000; Turrell et al. 2002; Wardle et al. 1997). Actually, criticism against animal foods is justified with the increase in saturated and all *trans*-fatty acids, the augment of added sugars and animal fats (Szpylka et al. 2018a). From the analytical angle, the interest in these topics (including also fructans because of their potential use against obesity) is increasing (Szpylka et al. 2018a, b). For these and other reasons, researchers have observed the increasing production of separated raw materials for the production of 'surrogates' able to mime the composition and nutritional supply of original animal foods (Carfì et al. 2018; González et al. 2015). At present, and with specific reference to vegan products, it can be assumed that the most important vegetable products favoured by active marketing campaigns are generally soy, coconuts, seeds, legumes, beans, and almonds (González et al. 2015; Seclì 2007).

The basic advantages of these productions—and vegan foods themselves—are (González et al. 2015):

(a) The absence of taste (for this reason, the food could exhibit different tastes because of the addition of a peculiar non-protein ingredient)
(b) Eco-friendly (eco-sustainable and socially acceptable) methods for agriculture and industrial transformation
(c) Lower prices if compared with animal protein sources
(d) The possibility of new and improved 'surrogate' products
(e) Improved technological features.

'Normal' consumers can choose to become vegans essentially for one or more of the following motivations (Larsson et al. 2003):

(1) Ethical reasons because of the need to avoid each possible pain to animals
(2) Repugnance for food of animal origin
(3) Health reasons because of the need to avoid animal fat and other compounds of animal origin (cholesterol intake is clearly one of the most recurring justifications)
(4) Explicit preference for vegetable/vegetarian foods and patterns if compared with animal products and complete dietary lifestyles
(5) Social pressure in terms of persuasion and media advertising has to be also considered in this ambit (Grbich 1990; Wentworth 1980).

It should be clarified that 'veganism' is the extreme vegetarian behaviour because of the complete exclusion of animal products, including also dairy products. In other terms, vegetarians exclude completely all possible types of mammal meats, meat preparations, poultry products, fish and seafood; milk and eggs are eliminated in spite of their 'non-living' status. Reasons are not scientifically justified, while vegetarianism (inclusion of animal products such as milk and eggs) and veganism should be seen as the product of philosophical tendencies (Larsson and Johansson 1997; Larsson et al. 2001).

On this basis, the above-mentioned question: 'could these animal imitation foods give the 'right' amount of proteins to vegan (and normal) consumers?' is fully justified. Actually, the same question can be considered when speaking of fatty acids, carbohydrates, or micronutrients. The problem of proteins should be discussed in terms of essential amino acids for the human being. On a general level, it might be considered that vegan-style diets have good effects on certain patients; although, only a few studies have been extensively carried out in this ambit and in relation to specific diseases only (Haddad et al. 1999; Hafström et al. 2001; Kjeldsen-Kragh et al. 1991, 1995; Nenonen et al. 1983).

4.2 Approximate Composition of Vegan Foods. Macroingredient Typologies

The approximate composition of vegan foods should be considered because of the need to replace all possible animal-derived components with adequate vegetable raw materials. For these reasons, a well-balanced recipe for each vegan typology should not exclude the following macrocategories of ingredients[1] (Belitz et al. 2009):

(1) Fat matters, e.g. refined palm oil. Refined palm oil is one of the most used vegetable materials for fat substitution (instant noodles, processed cheeses, etc.) and food-related operations such as industrial frying, after different refining processes. The main feature of this raw material is the ready solidification at low temperatures. Substantially, the original fat matter is extracted as a semisolid material and subsequently separated into two main fractions. Olein and refined fractions of this extract are extensively used in the food industry (Sect. 3.3). Different reasons, not only economic motivations, can justify the use of palm oil in many products, including technological improvements. By the health viewpoint, palm oil could diminish the risk of cardiovascular diseases, but the global increase of total (animal + vegetable) fats is a current worrying. However, this problem should be mitigated in pure vegetable foods

(2) Carbohydrates: modified polysaccharides such as agar, a gelatinous mass obtained from algae such as *Gelidium* spp., *Pterocladia* spp., *Gracilaria* spp., and certain *Rhodophyceae*. The mixture of polysaccharides, composed basically from β-D-galactopyranose and 3,6-anhydro-α-L-galactopyranose ($1 \rightarrow 4$ and $1 \rightarrow 3$ links are involved) makes possible the production of a three-dimensional network. Agar is essential as gelling agent; the mixture has to be obtained from the original polysaccharide by means of esterification (with sulphuric acid); the consequent mixture can be purified by gel congealing. Interestingly, the gelling action depends on the average approximate molecular weight of agar mixtures,

---

[1] One specific ingredient is used as a qualifying representative for the related category.

on the related concentration, and temperatures. In addition, it should be considered that gelified agars do not melt and gel at the same temperature or at the same intervals (Belitz et al. 2009)

(3) Proteins, e.g. soy proteins. These proteins (three categories: full-fat, defatted or lecithinated flour; concentrates; and isolates) have always represented a useful protein source in many developing countries. Obtained from soybeans (a legume type, of the family *Fabaceae*), these proteins can be extracted at pH 6.8 with good yields (80%). Soy proteins are usually subdivided into globulins (90% of the total protein amount), albumins (10%), and a negligible fraction of glutelins (Belitz et al. 2009). In relation to globulins, the glycine amount is prevailing (between 17.54 and 18.03%), followed by asparagine/aspartic acid (11.18–13.10%), while methionine is very low in all situations. It has to be noted that vitamin B_{12} (cobalamin) is not associated with soy proteins because of its localisation in animal proteins, although certain claims recommend the consumption of fermented soy products as the '*tempeh*' (Liem et al. 1977). Similarly to other legume proteins, soy proteins can act as emulsifiers, depending on pH value and temperatures at least.

4.3   Critical Evaluation of Vegan Foods. Are These Products Good Protein Sources?

At present, initial researches concerning veganism and vegetarianism reported that vegan consumers appear to rely on a sufficient protein intake, and the same thing appears correct with the exception of vitamin B_{12} and two minerals: zinc and iron (Bar-Sella et al. 1990; Dwyer 1991; Ellis and Montegriffo 1970; Ellis and Mumford 1967; Guggenheim et al. 1962; Harland and Peterson 1978; Latta and Liebman 1984; Lindenbaum et al. 1988; Sanders et al. 1978; Wokes et al. 1955). A recent study has also clarified that:

(a) Vegans are accustomed to having an increased—and expected—consumption of cereals, vegetables, legumes, fruits by vegans, nuts, and seeds. The remaining part of their diet relies on the consumption of soy proteins, tofu (a soy cheese) and meat analogues (produced with vegetable fats, gelling carbohydrates, and protein sources like soy protein, wheat gluten, etc.).

(b) In this specific ambit, it should be predicted—and it has been observed—that the energy intake related to proteins and fats is significantly lower when non-vegetarian people, especially when speaking of female subjects (male subjects show similar results).

(c) The supplementation of some deficient vitamin and/or metals (vitamin A, thiamine, folic acid, magnesium, iron, etc.) is useful with the exception of vitamin B_{12} at least (this vitamin is present only in animal foods).

(d) The intake of dietary fibres, zinc, and iron remains not satisfactory.

In relation to dietary fibres, other studies report higher consumption (Messina and Reed Mangels 2001; Waldmann et al. 2004). However, the same researches highlight potential risks, especially for children, in relation to riboflavin, vitamin D, zinc, calcium, and iodine. Anyway, it appears that the energy intake remains generally lower if compared with non-vegan lifestyles, suggesting the need for fortified vegan products. Consequently, the production of vegan-style foods has to consider this aspect and the deficiency of certain mineral substances strictly linked with animal proteins (a useful example is calcium, associated with caseins in milk, and virtually lower in vegan-style foods, on these bases). With specific reference to proteins, their possible deficiency can be solved by means of fortified recipes; the problem should not be considered only with the augment of daily portions because the assumption of fatty acids can dramatically increase. In exclusive relation to nitrogen balance and the need of sulphured amino acids, it has been reported that soy proteins could be satisfactory enough, although a methionine supplementation is preferable. A certain amount of vegetables, fruits, legumes, and cereals is certainly recommended when speaking of vegetable foods, including proteins, because of the lack of one or more essential amino acids at least in each of vegetable sources (Hoffman and Falvo 2004). Actually, the real need for supplementation is recommended for children and especially for newborns (Hoffman and Falvo 2004; Young 1991).

References

Bar-Sella P, Rakover Y, Ratner D (1990) Vitamin B12 and folate levels in long-term vegans. Isr J Med Sci 26(6):309–312

Belitz HD, Grosch W, Schieberle P (2009) Food chemistry, fourth revised and extended edition. Springer, Berlin

Carfì D, Donato A, Panuccio D (2018) A game theory coopetitive perspective for sustainability of global feeding: agreements among vegan and non-vegan food firms. In: Information Resources Management Association (ed) Game theory: breakthroughs in research and practice. IGI Global, Hershey, pp 71–104. https://doi.org/10.4018/978-1-5225-2594-3.ch004

Dwyer JT (1991) Nutritional consequences of vegetarianism. Annu Rev Nutr 11(1):61–91. https://doi.org/10.1146/annurev.nu.11.070191.000425

Ellis FR, Montegriffo VME (1970) Veganism, clinical findings and investigations. Am J Clin Nutr 23(3):249–255. https://doi.org/10.1093/ajcn/23.3.249

Ellis FR, Mumford P (1967) The nutritional status of vegans and vegetarians. Proc Nutr Soc 26(2):209–216. https://doi.org/10.1079/pns19670038

Elmadfa I, Freisling H (2005) Fat intake, dietvariety and health promotion. Diet In: Diversification and health promotion. Forum Nutr Basel, vol 57. Karger Publishers, Basel, pp 1–10. https://doi.org/10.1159/000083749

González A, Strumia MC, Igarzabal CIA (2015) Modification strategies of proteins for food packaging applications. In: Cirillo G, Spizzirri UG, Iemma F (eds) (2015) Functional polymers in food science. Wiley, Hoboken, and Scrivener Publishing LLC, Beverly, pp 127–145. https://doi.org/10.1002/9781119109785

Grbich C (1990) Socialisation and social change: a critique of three positions. Brit J Sociol 41(4):517–530. https://doi.org/10.2307/590665

Guggenheim K, Weiss Y, Fostick M (1962) Composition and nutritive value of diets consumed by strict vegetarians. Br J Nutr 16(1):467–471. https://doi.org/10.1079/bjn19620045

Haddad EH, Berk LS, Kettering JD, Hubbard RW, Peters WR (1999) Dietary intake and biochemical, hematologic, and immune status of vegans compared with nonvegetarians. Am J Clin Nutr 70(3):586S–593S. https://doi.org/10.1093/ajcn/70.3.586s

Hafström I, RingertzB, Spångberg A, von Zweigbergk L, Brannemark S, Nylander I, Rönnelid J, Laasonen L, Klareskog L (2001) A vegan diet free of gluten improves the signs and symptoms of rheumatoid arthritis: the effects on arthritis correlate with a reduction in antibodies to food antigens. Rheumatol 40(10):1175–1179. https://doi.org/10.1093/rheumatology/40.10.1175

Harland BF, Peterson M (1978) Nutritional status of lacto-ovo-vegetarian Trappist monks. J Am Diet Assoc 72(3):259–264

Hoffman JR, Falvo MJ (2004) Protein–which is best? J Sports Sci Med 3(3):118–130

Kaplan RM (1984) The connection between clinical health promotion and health status: a critical overview. Am Psycholog 39(7):755–765. https://doi.org/10.1037/0003-066x.39.7.755

Kjeldsen-Kragh J, Haugen M, Borchgrevink C, Laerum E, Haugen M, Eek M, Frre O, Mowinkel P, Hovi K (1991) Controlled trial of fasting and one-year vegetarian diet in rheumatoid arthritis. Lancet 338(8772):899–902. https://doi.org/10.1016/0140-6736(91)91770-u

Kjeldsen-Kragh J, Hvatum M, Haugen M, Førre O, Scott H (1995) Antibodies against dietary antigens in rheumatoid arthritis patients treated with fasting and a one-year vegetarian diet. Clin Exp Rheumatol 13(2):167–172

Larsson C, Johansson G (1997) Prevalence of vegetarians in Swedish secondary schools. Scand J Nutr/Næringsforskning 41:117–120

Larsson C, Klock K, Nordrehaug AÅ, Haugejorden O, Johansson G (2001) Food habits of young Swedish and Norwegian vegetarians and omnivores. Pub Health Nutr 4(5):1005–1014. https://doi.org/10.1079/phn2001167

Larsson CL, Rönnlund U, Johansson G, Dahlgren L (2003) Veganism as status passage. Appetite 41(1):61–67. https://doi.org/10.1016/s0195-6663(03)00045-x

Latta D, Liebman M (1984) Iron and zinc status of vegetarian and non-vegetarian males. Nutr Rep Int 30:141–149

Law M (2000) Plant sterol and stanol margarines and health. BMJ 320(7238):861–864. https://doi.org/10.1136/bmj.320.7238.861

Liem IT, Steinkraus KH, Cronk TC (1977) Production of vitamin B-12 in tempeh, a fermented soybean food. Appl Environ Microbiol 34(6):773–776

Lindenbaum J, Healton EB, Savage DG, Brust JCM, Garrett TJ, Podell ER, Margell PD, Stabler SP, Allen RH (1988) Neuropsychiatric disorders caused by cobalamin deficiency in the absence of anemia or macrocytosis. N Engl J Med 318(26):1720–1728. https://doi.org/10.1056/nejm198806303182604

Messina V, Reed Mangels A (2001) Considerations in planning vegan diets. J Am Diet Assoc 101(6):661–669. https://doi.org/10.1016/S0002-8223(01)00167-5

Nenonen MT, Helve TA, Rauma AL, Hänninen OO (1983) Uncooked lactobacilli-rich, vegan food and rheumatoid arthritis. Brit J Rheumatol 37(3):274–281. https://doi.org/10.1093/rheumatology/37.3.274

Sanders TAB, Ellis FR, Dickerson JWT (1978) Haematological studies on vegans. Br J Nutr 40(1):9–15. https://doi.org/10.1079/bjn19780089

Seclì R (2007) Veg-economy: fondamenti, realtà e prospettive dell'impresa 'senza crudeltà'. Dissertation. Libera Università Internazionale degli Studi Sociali 'Guido Carli', Rome

Szpylka J, Thiex N, Acevedo B, Albizu A, Angrish P, Austin S, Bach Knudsen KE, Barber CA, Berg D, Bhandari SD, Bienvenue A, Cahill K, Caldwell J, Campargue C, Cho F, Collison MW, Cornaggia C, Cruijsen H, Das M, De Vreeze M, Deutz I, Donelson J, Dubois A, Duchateau GS, Duchateau L, Ellingson D, Gandhi J, Gottsleben F, Hache J, Hagood G, Hamad M, Haselberger PA, Hektor T, Hoefling R, Holroyd S, Holt DL, Horst JG, Ivory R, Jaureguibeitia A, Jennens M,

Kavolis DC, Kock L, Konings EJM, Krepich S, Krueger DA, Lacorn M, Lassitter CL, Lee S, Li H, Liu A, Liu K, Lusiak BD, Lynch E, Mastovska K, McCleary BV, Mercier GM, Metra PL, Monti L, Moscoso CJ, Narayanan H, Parisi S, Perinello G, Phillips MM, Pyatt S, Raessler M, Reimann LM, Rimmer CA, Rodriguez A, Romano J, Salleres S, Sliwinski M, Smyth G, Stanley K, Steegmans M, Suzuki H, Swartout K, Tahiri N, Ten Eyck R, Torres Rodriguez MG, Van Slate J, Van Soest PJ, Vennard T, Vidal R, Hedegaard RSV, Vrasidas I, Vrasidas Y, Walford S, Wehling P, Winkler P, Winter R, Wirthwine B, Wolfe D, Wood L, Woollard DC, Yadlapalli S, Yan X, Yang J, Yang Z, Zhao G (2018a) Standard Method Performance Requirements (SMPRs®) 2018.001: sugars in animal feed, pet food, and human food. J AOAC Int 101(4):1280–1282. https://doi.org/10.5740/jaoacint.smpr2018.001

Szpylka J, Thiex N, Acevedo B, Albizu A, Angrish P, Austin S, Bach Knudsen KE, Barber CA, Berg D, Bhandari SD, Bienvenue A, Cahill K, Caldwell J, Campargue C, Cho F, Collison MW, Cornaggia C, Cruijsen H, Das M, De Vreeze M, Deutz I, Donelson J, Dubois A, Duchateau GS, Duchateau L, Ellingson D, Gandhi J, Gottsleben F, Hache J, Hagood G, Hamad M, Haselberger PA, Hektor T, Hoefling R, Holroyd S, Lloyd Holt D, Horst JG, Ivory R, Jaureguibeitia A, Jennens M, Kavolis DC, Kock L, Konings EJM, Krepich S, Krueger DA, Lacorn M, Lassitter CL, Lee S, Li H, Liu A, Liu K, Lusiak BD, Lynch E, Mastovska K, McCleary BV, Mercier GM, Metra PL, Monti L, Moscoso CJ, Narayanan H, Parisi S, Perinello G, Phillips MM, Pyatt S, Raessler M, Reimann LM, Rimmer CA, Rodriguez A, Romano J, Salleres S, Sliwinski M, Smyth G, Stanley K, Steegmans M, Suzuki H, Swartout K, Tahiri N, Eyck RT, Torres Rodriguez MG, Van Slate J, Van Soest PJ, Vennard T, Vidal R, Vinbord Hedegaard RS, Vrasidas I, Vrasidas Y, Walford S, Wehling P, Winkler P, Winter R, Wirthwine B, Wolfe D, Wood L, Woollard DC, Yadlapalli S, Yan X, Yang J, Yang Z, Zhao G. (2018b) Standard Method Performance Requirements (SMPRs®) 2018.002: fructans in animal food (Animal feed, pet food, and ingredients). J AOAC Int 101(4):1283–1284. https://doi.org/10.5740/jaoacint.smpr2018.002

Turrell G, Hewitt B, Patterson C, Oldenburg B, Gould T (2002) Socioeconomic differences in food purchasing behaviour and suggested implications for diet-related health promotion. J Hum Nutr Diet 15(5):355–364. https://doi.org/10.1046/j.1365-277x.2002.00384.x

Waldmann A, Koschizke JW, Leitzmann C, Hahn A (2004) Dietary iron intake and iron status of German female vegans: results of the German vegan study. Ann Nutr Metabol 48(2):103–108. https://doi.org/10.1159/000077045

Wardle J, Steptoe A, Bellisle F, Davou B, Reschke K, Lappalainen R, Fredrikson M (1997) Health dietary practices among European students. Health Psychol 16(5):443–450. https://doi.org/10.1037/0278-6133.16.5.443

Wentworth WM (1980) Context and understanding—an inquiry into socialization theory. Elsevier, New York

Wokes F, Badenoch J, Sinclair HM (1955) Human dietary deficiency of vitamin B12. Am J Clin Nutr 3(5):375–382. https://doi.org/10.1093/ajcn/3.5.375

Young VR (1991) Soy protein in relation to human protein and amino acid nutrition. J Am Diet Assoc 91(7):828–835